CONTENTS

Maps x
Introduction 1

Chapter 1: Pioneers · 13

1 Step Pyramid of Djoser 15
2 Temple of Apollo Epicurius 19
3 The Pantheon 24
4 Palazzo Rucellai 27
5 Villa Almerico Capra 32
6 Château de Vaux-le-Vicomte 35
7 Marshall, Benyon, and Bage's Flax Mill 39
8 Dulwich Picture Gallery 41
9 The Palm House 45
10 E. V. Haughwout Building 48
11 Oriel Chambers 51
12 St Pancras Station 55
13 Guaranty Building 61
14 Flatiron Building 64
15 Woolworth Building 69
16 AEG Turbine Hall 72
17 The Steiner House 74
18 The Rietveld Schröder House 77
19 The Bauhaus 81
20 The Lovell Health House 85
21 Narkomfin Apartment Building 89
22 La Maison de Verre 91

23 Chrysler Building 94
24 Villa Savoye 98
25 Salginatobel Bridge 101
26 The Melnikov House 104
27 Seagram Building 107

Chapter 2: Rhetoric   111

28 Tower of the Winds 113
29 Crac des Chevaliers 116
30 Alhambra 120
31 Rushton Triangular Lodge 123
32 Old Shoin and The New Palace 126
33 Easton Neston 131
34 Bhutanese Farmhouse 135
35 Altes Museum 138
36 Thorvaldsen Museum 141
37 Wharenui 144
38 Stockholm Public Library 147
39 Fallingwater 150
40 Casa Malaparte 154
41 Säynätsalo Town Hall 157
42 National Parliament 160
43 Trellick Tower 163

Chapter 3: Sacred   167

44 Mortuary Temple of Hatshepsut 169
45 Temple of Solomon 172
46 St Catherine's Monastery 177
47 Hagia Sophia 180
48 Dome of the Rock 184
49 Malwiya 187
50 Mausoleum of Isma'il Samani 190
51 Chapel of St John 193
52 Durham Cathedral 196

53 Chartres Cathedral 200
54 Minaret of Jam 203
55 Angkor Thom 206
56 Rock-Cut Churches 209
57 Sri Ranganathaswamy Temple 212
58 Janggyeong Panjeon 215
59 Basilica of San Lorenzo 218
60 Santa Maria dei Miracoli 221
61 Hanging Temple 223
62 The New Sacristy 226
63 Imam Mosque 231
64 Taj Mahal 234
65 Temple Expiatori de la Sagrada Familia 237
66 Great Mosque 240
67 Church of the Sacred Heart 243
68 Notre Dame du Haut 246

Chapter 4: Urban Visions   251

69 Uruk 253
70 Persepolis 256
71 Petra 260
72 Leptis Magna 264
73 Palmyra 267
74 Palenque 271
75 Machu Picchu 274
76 Place des Vosges 277
77 The Circus 280
78 Sydney Opera House 286

Chapter 5: Big And Beautiful   291

79 Great Pyramid at Giza 293
80 Great Wall of China 298
81 Taq-i Kisra 301
82 Forbidden City 304

Chapter 6: Material Matters   309

83   The Barley Barn 311
84   Ca' d'oro 315
85   Shibam 318
86   Himeji Castle 321
87   Church of the Transfiguration 325
88   Beurs van Berlage. 326
89   The Majolikahaus. 330
90   Marshcourt 333
91   Art Gallery of Ontario. 336
92   Farnsworth House 340
93   Hopkins House 343
94   Lloyd's of London 348
95   Rozak House 350

Chapter 7: Lost And Found   355

96   Ishtar Gate 357
97   Frauenkirche 360
98   Catherine Palace. 363
99   Euston Arch 366
100   Hatra 370

Glossary 377
References 381
Bibliography 384
Index 389

LOCATION OF BUILDINGS: Map 1

LOCATION OF BUILDINGS: Map 2

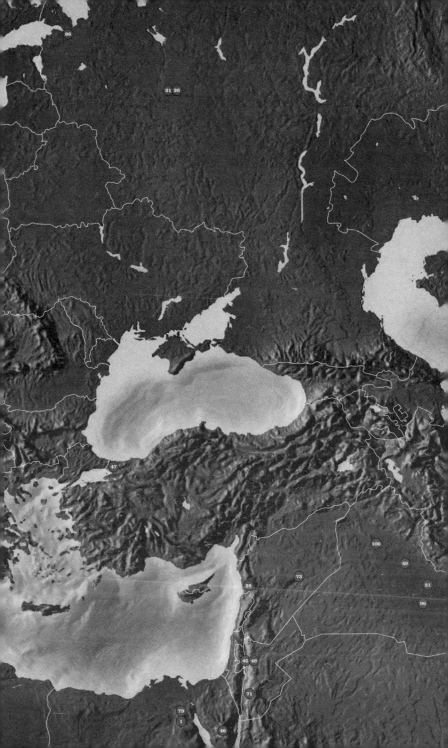

INTRODUCTION

THIS BOOK OFFERS A history of world architecture through the stories of 100 buildings. The selection includes many of the world's best-known buildings that represent key or pioneering moments in architectural history, but there are also less obvious and more humble structures, the generally unsung heroes of a great and fascinating story. Virtually all the buildings included are ones that I've seen – this has been one of the criteria of selection – so the story this book tells is a most personal and intimate affair. In addition, buildings have been selected to represent most major building types and a wide geographic spread that reflects the architectural aspirations of many cultures through many epochs.

By its nature a history of architecture is also a history of many other things – of politics, economy, society, religion, science, and ecology, and of art and culture generally – and so is essentially a history of the world. For architecture, in its forms and functions, is a very direct mirror of mankind's desires, concerns, and ambitions.

Architecture has created bastions of defence and aggression, homes for the gods and the dead; temples of commerce

1

and of the arts; palaces to express taste, power, and wealth; and shrines of science, of learning, knowledge, and politics – but also of Mammon and of physical and spiritual subjugation and incarceration. Can an ignoble cause create great architecture? Can a great fortress – perhaps a place of oppression – be a thing of beauty as well as of power? These are perennial philosophical questions that this book seeks to answer.

The book also explores moments of human madness when culture comes under fire. During war, and because of ideological, political, or religious conflict, buildings – particularly cities – have been targets, sometimes victims of collateral damage and sometimes of direct and brutal attack. This book documents this dark side of the story of architecture, referring to examples from history and from the present, including in Iraq, where in March 2015 architecture and art of great age and beauty were deliberately destroyed – a stark and tragic sacrifice to mankind's ability to lose itself in intolerance and unbridled fury. These recent losses – notably the attacks on the cultural treasures of the ancient city sites of Nineveh, Nimrud, and Hatra, long regarded as among the world's most precious antiquities – make clear that in our increasingly hostile, volatile, and divided, world nothing is safe.

Together, the stories told in this book offer a global cultural history – a history that is still in part mysterious and ripe for informed speculation. Some of the basic facts to do with the emergence, construction, and evolution of these edifices remain unknown or contested and obscured by ancient myth and legend. How was the Great Pyramid at Giza designed and constructed with such breathtaking perfection and accuracy around 4,550 years ago? In what circumstances did the classical architecture of the Greek world reach maturity nearly 2,600 years ago? By what leap of imagination and engineering skill did the revolutionary system of Gothic architecture suddenly emerge – in fully formed manner – in the early twelfth century?

The rise of the Gothic is a fantastic story. What is its relationship with contemporary Islamic architecture of the Middle East and Southern Spain, and was the grain of the Gothic idea gathered during the Crusades and brought back to Western Europe?

In a generation in Europe in the very early twelfth century, all structural traditions were dramatically revised and an integrated system of skeletal architecture comprising ribs, piers, and buttresses replaced the tried and trusted tradition of wall architecture, in which loads were transferred and restrained through the means of walls of great strength. Suddenly architects, masons, and engineers grasped the potential of strength not through mass but by design. Materials could be given added strength if used in a well-calculated manner and structures of great strength could be minimal if all was conceived in harmony, with the thrust of forces being met and balanced by opposite and equal counter-thrusts – for example, the lateral thrust of the weight of a stone vault being met with the counter-thrust of a well-placed flying buttress. This was an architecture that largely relieved the wall of its structural and load-bearing role, freeing it to be pierced by huge windows that, filled with stained glass, allowed God's light and message to flood into the interiors of churches and cathedrals.

Even the very definition of architecture, as opposed to mere building, has been debated and honed for generations.

The Roman architect and theorist Vitruvius, in the first century BCE, offered the oldest known definition: that architecture must embody 'commodity, firmness, and delight'. In simple terms this means that architecture must be functional and fit for purpose (in other words, commodious), and that it must be structurally sensible and sound, and so possess 'firmness'. But for mere building to become architecture it must also attain the poetry of beauty; it must possess a meaning and a spiritual quality that feeds and inflames the imagination, that satisfies human fancies – that delights.

The greatest mystery, perhaps, is how and why some basic forms and details in architecture spread around the globe in ancient times to become the universal expressions of many of mankind's most fundamental beliefs. For example, why do pyramids, as sacred forms and as places often associated with death and rebirth, appear in different cultures around the world – in Mesopotamia, in Mayan architecture in Central and South America, in Sri Lanka?

Where lies the origin of the city – in the Middle East, in Central Asia? What is the true history of the humble brick, the most ancient and universal of man-made building materials, which possesses both great strength and beauty and when fired is a pure act of alchemy by which the base elements of earth and water are transformed into a material that can endure through the centuries, that is stronger than stone? The brick is embedded in the cultural history of the Judaic, Christian, and Islamic religions because the process of brick-making and the difficulties involved did much to provoke Moses to lead the Children of Israel out of bondage in Egypt to the Promised Land. As explained in the Book of Exodus, the pharaoh, to increase the persecution and labour of the Israelites, refused to give them straw to make their bricks yet demanded that they meet their daily quota 'as when there was straw'. The Israelites were compelled to scatter to 'gather stubble instead of straw' (Exodus 5: 12–14), and then to flee. The consequences of the flight that this persecution provoked still echo through history. And the world's oldest book – *The Epic of Gilgamesh*, written perhaps 4,700 years ago – extols the virtue of the brick. Gilgamesh sought immortality and eventually found it in architecture, through the construction of city walls wrought of hard, kiln-fired bricks stamped with his name – as 'the names of famous men are written'. The walls were to endure for eternity, with the memory of the king's name kept alive within them, stamped on their bricks.

Gilgamesh was the king of one of the world's first cities – Uruk in Mesopotamia – and one of the places where civilization as we conceive it started nearly 5,500 years ago. I have been to Uruk and seen for myself the quality of its kiln-fired bricks and clay cones – used as a weather-proof veneer over more fragile sun-dried bricks – that have lasted, beautiful and intact, through the millennia. And I've seen the names of ancient Mesopotamian kings stamped on bricks and cones – of Naram-Sin, a king of Uruk around 3,900 years ago, and of Nebuchadnezzar II at Babylon and Borsippa in Iraq, where he ruled 2,600 years ago.

And when in Uruk, I discovered another architectural mystery – or perhaps a solution to a mystery – in the fragments of clay-cone-clad walls of sacred or royal buildings that emerged above the dry and sand-covered ground. These fragments of wall were constructed around 5,000 years ago and are pure architecture in the Vitruvian sense, for their kiln-fired cones not only offered 'firmness' by providing practical protection for the softer sun-dried bricks behind, but also offer beauty – and beauty of a most baffling kind. The cones are of different colours and are arranged to create abstract surface patterns – notably spirals, lozenges, and chevrons. When I first saw these patterns, emerging out of the dust of the ancient city, I was stunned. These are the very patterns, enigmatic and provoking, that I have seen on medieval mosques and minarets in the Middle East and in Western European churches and cathedrals of the late eleventh and early twelfth centuries – notably on the vast transept and nave columns of Durham Cathedral. These mysterious patterns are evidently the ancient language of the divine, of the sacred. Their spread over great distances of time and space suggests how recurring architectural images and forms inspired the ancient world, presumably observed and copied by merchants, warriors, and clerics travelling the great trade and pilgrimage routes along which passed not just goods but cultural and religious ideas.

This book explores not just architectural connections but also the fundamental differences between cultures around the world. In the West there has evolved an emphasis on the physical, material solidity and permanence of building, expressed through robust materials and sound engineering – an inheritance, perhaps, of the ziggurat and pyramid builders. In the East, however, the spiritual rather than material aspects of architecture are more important. The ideas or beliefs that a building embodies are more important than the physical reality of the building itself. So lightness of construction – timber and unfired earth or mud rather than stone or brick – characterize architecture in Japan or Bhutan, where an ancient building is not one that expresses a continuity of material existence but one – perhaps often rebuilt many times – that continues an idea or belief, a function, or the use of a time-hallowed sacred site; an architecture that reflects the Buddhist doctrine that all the material world is transitory, that the only certainty is that there is no certainty but only change, and that it is the spiritual not the material that endures.

The philosophical propositions that underpin much architecture are explored throughout the book – notably that belief in the application of 'divine' systems of harmonious proportion by Greek, Roman, and Renaissance designers, and the conviction that truth (for example, in the expression of materials and means of construction) will produce beauty. The engineering aspect of architecture is also investigated: for example, the project to build vast domes and arches – forms of great strength but which are demanding in their execution – that characterizes Roman construction and which becomes an obsession in late Persian, Byzantine, Islamic, and Renaissance architecture.

The exploration of the engineering aspect of architecture is pursued around the globe – for example, in the examination of early mosques and of vast Hindu temples in south India, where cave-like spaces are almost excavated out of, rather than constructed within, pyramid-like masonry towers and *gopuram* gates. The story of engineering also characterizes much of Western architecture during the nineteenth century, when new methods and materials revolutionized architecture, making new types and scales of building possible – for example, cast- and wrought-iron train sheds of huge span, and ultimately iron- and then steel-framed high-rise architecture, pioneered in late eighteenth-century Britain during the Industrial Revolution and realized in its full potential in the late nineteenth and early twentieth centuries in Chicago and New York.

This story of nineteenth-century technological progress in Europe and the United States is set within the cultural ethos of an age when, often in a baffling manner, even the most modern buildings were given the pedigree of history, a veneer of the past to make them artistically acceptable. So railway sheds were given Gothic detail – as at Brunel's Paddington Station in the early 1850s – to imply an analogy with medieval cathedrals. And modern building types, such as large hotels and public buildings incorporating a modern mix of uses and services, were designed to look like Gothic monastic buildings and guild halls, or like Renaissance palazzi, as in the case of George Gilbert Scott's Foreign Office in Whitehall. The results can be charming, if a trifle illogical.

Engineering and the application of new materials and systems of construction also define much of the story of architecture in the twentieth and twenty-first centuries. The application of steel-reinforced concrete transformed design, construction, and the aesthetics as well as the economy of architecture, but brought artistic and functional ailments along with its advantages. Structural steelwork has also changed the nature of architecture, and allowed it to achieve the new building types needed to satisfy the unprecedented demands of the modern age, including high-rise housing and commercial buildings.

But as well as new building technologies, the last hundred years have seen new architectural concepts and theories – some progressive and benign, others troubling and malign. There have been attempts to create modern utopias that have been little more than misguided experiments in social engineering, leaving a legacy of mismanaged high-rise housing. Some of these schemes have involved economically driven forms of system-building that have ended in catastrophe, as seen in the 1968 partial collapse of the twenty-two-storey Ronan Point building in East London. By excluding the concept of 'delight', much modern system-building has reduced architecture to no more than mere building.

Other experiments have been more intriguing – some even more successful. British 'High Tech' architecture of the 1970s and 80s combined new material and methods of construction with revolutionary concepts. It viewed architecture as an essentially functional and disposable artefact – much like a car or fridge – that should utilize standard components, have its more transient elements (such as services and lifts) located on the exterior for easy replacement, be flexible in plan, and be disposed of like any other modern machine when no longer useful. Piano and Rogers' Pompidou Centre in Paris and Rogers' Lloyd's Building in London are famous expressions of these ideas. Ironically – indeed paradoxically – both of these functional machines, built for disposal, are now regarded officially as cultural icons and are protected from change.

More significant in its long-term and wide-reaching influence has been the sculptural concrete architecture that evolved in the Soviet Union in the 1920s, and which became known as Constructivism. This movement gave birth to a twentieth-century architecture – Beton Brut in France, Brutalism in Britain – that flourished from the late 1940s, possesses a certain rough and rude poetic beauty – based on the concept of 'truth to materials' – and has produced such homegrown examples as Ernö Goldfinger's Trellick Tower in West London (admittedly more Constructivist than Brutalist), Denys Lasdun's National Theatre, and buildings in Bangladesh and the United States by Louis Kahn. Some of the most moving examples were produced by Le Corbusier in the 1940s, 1950s, and 1960s, such as the Unité d'Habitation housing project in Marseille.

Constructivism was an art movement with a political and social message. It was 'socially progressive' and presented architecture as an expression and means of change. It pursued the notion that architecture has a moral purpose beyond the simple world of aesthetics. In the 1830s, the English architect and polemicist A W N Pugin argued that there was a right and a wrong way to build. Classical design was wrong because it was pagan and foreign in origin and structurally primitive. Gothic design was right because it was Christian in inspiration, had been given a native British character, and was architecturally and structurally superior to classical architecture, with a language that was more appropriate to the modern age. So Pugin brought an aspect of morality into architecture, but one trammelled by the concerns of art, national identity, and theology. Constructivism went beyond this and took architecture into the wide and sweeping world of political ideology and social purpose.

Constructivist architecture was a declaration of belief and of intent. It also reinforced the notion that architecture can bring about change or safeguard treasured values. Battle lines have been drawn and separate camps established that, in their products, now define much of Western architecture. There are 'progressive' architects and their clients who promote 'modern' design, and, opposed to them, there is the camp that seeks to conserve and enhance the past, to take inspiration from history and to design and build in a traditional manner. But uniting these two camps is an increasing concern for non-polluting, sustainable, and energy-conscious ecological or 'green' architecture that demonstrates the contribution building can make towards saving, rather than harming, the planet.

Architecture is an all-embracing adventure without end. It is a story that can never be completed as long as mankind continues to build, to invent, to discover; it is the story told by this book.

CHAPTER 1

PIONEERS

SEEN THROUGH THE PRISM of invention and discovery, even the most ancient buildings can seem as new. The Step Pyramid at Saqqara, Egypt, is perhaps nearly 4,700 years old and now battered by time. But it was from this structure that the pyramid evolved in Egypt – possibly anticipated by ziggurats in Mesopotamia – leading quickly to the sloping-sided pyramids at Giza, which remain among the most astonishing architectural creations on earth. The step pyramid also seems to have pioneered the use of stone as a building material in Egypt. Although used in quite a crude manner at Saqqara, stone was soon used to achieve immaculate construction, incorporating huge blocks cut with incredible precision, to become one of the wonders of Egyptian architecture.

The Temple of Apollo Epicurius at Bassae, Greece, around 2,400 years old, possesses the clarity of conception and execution that has made Greek architecture an enduring touchstone of excellence, while the Pantheon in Rome, now nearly 2,000 years old, continues to astonish with the power of its design, the audacity of its scale, and the innovation of its technology of construction.

Innovation, combined with a quest for ever greater scale, characterizes much of the pioneering architecture of the late eighteenth to twentieth centuries. Marshall, Benyon & Bage's Flax Mill, in Ditherington, England, built in 1796, capitalized on the technological inventions of the Industrial Revolution to create a largely iron-framed, brick, and supposedly 'fire-proof' structure that helped lay the foundations for later iron-framed structures. These include the Palm House, Kew Gardens, London, of 1844–48 and the train shed of 1865–68 at St Pancras Station, London. They combined wrought-iron with cast iron, and introduced large areas of glass to create minimal structures of great beauty and strength.

This iron-framed building technology, later realized in steel or reinforced concrete and combined with other developments such as the reliable elevator and fire-proof cladding materials, led to the great novel urban building type of the late nineteenth century – the skyscraper. One the earliest is Sullivan and Adler's Guaranty Building of 1896 in Buffalo, USA, which rises thirteen storeys, its steel frame clad with incombustible terracotta. Realizing the artistic potential of the building type is Mies van der Rohe's Seagram Building, New York, of 1958 which, with its generously glazed curtain wall and minimal structure, remains one of the most perfectly conceived high-rise towers ever built.

1

STEP PYRAMID OF DJOSER
Saqqara, Egypt | c. 2650 BCE

The Step Pyramid at Saqqara is the most visually dominant feature at the heart of an ancient necropolis. It is also generally assumed to be the first of the pyramid structures in Egypt to be constructed of stone, and so is an epoch-making pioneer.

Egypt did not produce the world's earliest monumental architecture; this appears to have evolved in Mesopotamia some centuries earlier, and the stone circles in Northern Europe – some, like Stonehenge, dating back perhaps as far as 3000 BCE – are, intellectually and structurally, self-evidently works of architecture. But Egyptian architecture initiated a chain of consistent development that, in the complexity of its conception and the excellence of its construction, inspired not just later epochs of the ancient world, notably Greece and then Rome, but also the European Renaissance. And, arguably, this development of Egyptian monumental architecture started with the Step Pyramid at Saqqara, which the consensus of informed opinion now suggests is perhaps a century older than the larger and far more sophisticated Great Pyramid at Giza.

The use of stone as the prime construction material at Saqqara was a tremendous innovation in a region where brick – sun-dried or kiln-fired – had been the usual building material and was to remain so for many buildings for centuries to follow. Brick's dominance is suggested most strongly by Old

Testament texts, perhaps 3,500 years old, which make it clear that brick-making was big and important business. The reason for the use of stone rather than brick is, at one level, entirely practical. Good-quality building stone – limestone and granite – was generally available in large parts of Egypt, while it was not usually available in large parts of Mesopotamia. But also, it would seem that in the ancient world specific stones were seen as possessing distinct qualities – not just structural qualities, but also symbolic and ritualistic ones. Different 'bluestones', notably Preseli spotted dolorite from Wales, and sandstone 'sarsens' were used at Stonehenge; the reason why is not clear, but their use would have involved significant efforts during construction. And the Great Pyramid at Giza incorporates at least two different types of limestone and granite from Aswan, although this use of different stones was more certainly for structural and aesthetic reasons.

The construction of the Step Pyramid of Djoser appears to have been an experimental affair, with its form evolving throughout the process. The relationship between its final step-form and the perhaps contemporary stepped or terraced ziggurats of Sumeria remains uncertain; even their chrono-logical relationship is unclear, although it seems likely that the structure at Saqqara, which eventually took the form of a step pyramid, slightly predates most surviving Sumerian ziggurats, with the probable exception of that at Uruk.

Although there are similarities between the Step Pyramid and ziggurats, there are also two clear and fundamental differences. First, of course, is the fact that ziggurats are made of sun-dried brick clad with hard, kiln-fired brick, not stone – although, no doubt significantly, the stone blocks used at

Saqqara are relatively small (certainly in comparison to the huge blocks used later at the Great Pyramid), so they could be handled much like bricks. Again, this can be read as evidence of the Step Pyramid's experimental or transitional status in the development of stone building technology.

But more significant perhaps is the fact that ziggurats were almost certainly temples, topped with shrines, and were not constructed as tombs (none have been found to contain obvious burial chambers). The pyramids of Egypt, on the other hand, appear to have served as, or been very closely connected with, royal burial and commemoration.

The Step Pyramid appears a most pragmatic affair and is almost certainly a tomb in origin. Informed speculation and interpretation of evidence suggest that it was designed by the polymath Imhotep, vizier to the pharaoh Djoser, evidently an architect and later deified as a magician, high priest, and healer. (Imhotep is named on a statue of Djoser.) Archaeological evidence suggests convincingly that the tomb's first tenant was Djoser himself, who is now known as one of the earliest and most important pharaohs of Egypt's Old Kingdom. It is highly likely that Djoser had a profound personal influence on the development of Egyptian stone-built architecture, for according to surviving fragments of *Aegyptiaca* (*History of Egypt*), written in the third century BCE by the Egyptian historian and priest Manetho, it was Djoser (or Tosorthros) who 'discovered' how to build with hewn stone.[1]

The process of the construction of the Step Pyramid is uncertain, but it seems that Imhotep first constructed a burial chamber and a labyrinth of passages and then, in line with convention, raised above these works a horizontal slab or *mastaba* – built, in pioneering fashion, of stone rather than brick. Then, for some critical but now unknown reason, a process was set in motion that led to the transformation of a simple *mastaba* tomb into a structure far more complex in form and meaning. The tomb was increased in width and raised in height by superimposing five more *mastabas* on the original, but each decreasing in size, to create a six-step pyramid. The core of roughly squared stones, which is pretty well all that now survives, was originally faced with finely cut limestone blocks, suggesting that enormous revenues must have been spent perfecting this vast and pioneering masonry structure. The meaning of the design that evolved remains shrouded in mystery, but evidently in Egypt 4,600 years ago tombs were more than just repositories for bodies and memories – they were also places of protection and machines for the rebirth of the soul. The stepped form might have represented a stairway to heaven, it might have symbolized the stages in the journey of the soul after death, or it perhaps had a role that was even more esoteric. What is certain is that, on the north side of the pyramid, a small mortuary chapel or *serdab* was constructed, within which was placed a powerfully wrought image of Djoser. This statue provided an earthly refuge for the *ka* – an aspect of the soul representing the spirit and essential life force – of the dead pharaoh, who could peer for eternity through a tiny porthole cut in the wall of the chapel. And the object of his unblinking gaze was the constant constellation in the northern hemisphere with Thuban – the Pole Star, 4,500 years ago – as its beacon. This constellation was deemed the home of the gods, and the image of the deified Djoser was doing no more than gazing at the place from which his soul had come and to which it hoped to return.

2

TEMPLE OF APOLLO EPICURIUS
Bassae, Andritsaina, Peloponnese, Greece
| c. 429–400 BCE

The evolution of the artistic perfection that Greek sacred and public architecture achieved by the fifth century BCE remains a subject of intense debate and of some mystery. This debate stretches back over 2,000 years, with, for example, the Roman architect Vitruvius attempting to solve the mystery by offering legends as explanations for the origin of Grecian architecture. One of the more quaint of Vitruvius' explanations touches on the birth of the ornamental Corinthian capital. Vitruvius explains that the capital was devised by Callimachus, who was delighted by the sight of the leaves and tendrils of an acanthus plant growing around a basket placed on a tomb. Inspired, writes Vitruvius, Callimachus immediately designed the Corinthian capital.[2]

In the eighteenth century, the origin of Greek architecture was explored in a more rational and systematic manner by such theorists as Marc-Antoine Laugier, who, picking up clues from Vitruvius, argued in his 1755 *Essay on Architecture* that stone-built Greek temples were inspired by natural or primitive structures. For many in the late eighteenth century, the 'primitive hut' – with tree trunks as columns and a simple pitched roof suggesting pediment and entablature – was seen to represent the elemental origins of classical architecture. This theory was simplistic, but it is generally accepted that Greek stone-built architecture evolved out of timber construction because many purely ornamental details of Greek masonry architecture can be seen as the functional expression of timber construction – for example, triglyphs in a Doric frieze arguably represent the ends of three planks joined together to form a structural beam, while the triangular or conical *guttae* beneath the trilgyph can be seen as memorials of the timber pegs that held the beam in place. Certainly, the post-and-lintel structural principle of Greek stone-built architecture reflects the natural constraints of timber building. But in the last 200 years or so it has become increasingly clear that Greek architecture, in its forms and proportions, was deeply dependent on Egyptian precedent. For example, the fluted columns in the outer court of Hatshepsut's Mortuary Temple near Luxor – around a thousand years older than the architecture of Greece's Golden Age – anticipated, perhaps even inspired, the shafts of Grecian Doric columns.

Combined with the excellence of construction that characterizes Greek architecture in the fifth century BCE is an awareness of the power of proportion. Again, this is touched on by Vitruvius, who essentially roots proportion in the observation and imitation of the natural world – the creation of the gods – and in particular in the differing proportions of the male and female body. The Greeks, stated Vitruvius, wanting to build columns that were strong and that also had a 'satisfactory beauty of appearance', compared the length of a man's foot with his height. Discovering that this gave a ratio of 6:1, 'they applied the same principle to the column, and reared the shaft, including the capital, to a height six times its thickness at its base'. Thus, explains Vitruvius, was the Doric column born, with 'the proportions, strength, and beauty of the body of a man'. In the same manner, the more slender, more beautiful and more feminine Ionic order was devised, but with its proportion based on the relationship between the length of a woman's foot and her height, leading to a 7:1 ratio between column height and diameter. Vitruvius explained that the ornamental Ionic capital had volutes like 'curly ringlets', and the column shaft had flutes 'like the folds in the robes worn by matrons'. So while the Doric order possessed manly beauty, 'naked and unadorned', the Ionic had 'the delicacy, adornment and proportions characteristic of women'. The third Greek order, the Corinthian, more ornamental and delicate than the matronly Ionic, was, according to Vitruvius, proportioned in 'imitation of the slenderness of a maiden'.[3]

The Temple of Apollo Epicurius, which stands in lofty splendour and isolation on the edge of Mount Kotilion in the southwest corner of ancient Arkadia, offers fascinating insights into the evolution of the three orders of classical architecture. It is currently believed that the temple was built by the Arkadian city-state of Phigalia, and meticulous and comparative analysis of the ruins suggests a construction date of c.429–400 BCE. The site is magnificent, and suggests that this is a building raised by a people who venerated divine creation – landscape, plants, the rising and setting of the sun – and who wanted to learn and apply the secrets of natural beauty. The Greek traveller Pausanias, writing in the second century CE, praised the temple for the beauty of its construction and design and stated that it was the work of Ictinus, one of the architects of the Parthenon in Athens. Few now agree with this attribution.

When the temple was first fully documented by Western archaeologists in 1811 and 1812, it caused a sensation because it broke many of the rules that, it was assumed, characterized Greek temple building. It is oriented north–south instead of the far more usual east–west, with its entrance front on the north, not the east. The proportion of its plan is peculiarly long and thin. It measures 14.5 x 38.3 metres (47.5 x 125.7 ft) from the top step of the stylobate, so the plan is of 2:5 proportion. The plans of rectangular Greek temples of the period do vary. The stylobate of the Parthenon in Athens (447–436 BCE) is just over 2:1, and that of the Temple of Aphaia on Aegina (500 BCE) just under 2:1; the 'Basilica' (530 BCE) at Paestum is just about 2:1, while the 'Temple of Neptune' at Paestum (460 BCE) is almost 2:5 and the Olympieion at Athens (515 BCE to 129 CE) is just over 2:5.

All these temples are in the Doric order (with the exception of the Olympieion, which is Corinthian), with rows of columns surrounding the naos. Outwardly the Temple of Apollo Epicurius is also Doric, with six columns along its short sides and fifteen visible along its long sides. All the columns, like the walls, are wrought of local blue-grey limestone.

Inside the temple, however, there are two great surprises. The naos contains Ionic columns of distinct and original design, with bases that flare out and volutes that are large, flat, and curvaceous – but most curious of all, the columns are attached to the naos by means of spur walls. So columns and walls are visually and structurally united in unprecedented manner to create a series of shallow but discreet recesses or shrines within the naos. But even this arrangement is not the greatest surprise. At the south end of the naos is a single central column – this is odd in itself, since Greek practice was generally to have a space rather than a column in the centre of a temple composition (though there are exceptions such as the 'Basilica' at Paestum in Italy), but odder still is the fact that the capital of the column was proto-Corinthian. Until this temple was analysed it was assumed that an individual Greek temple would be designed consistently in one of the three Greek orders – Doric, Ionic, or Corinthian – and not embody elements of all three.

The Ionic columns and the isolated Corinthian column all helped to carry the same 'wrap-around' frieze, and so were united in their function. The frieze is made of imported marble and depicts Greeks and Herakles fighting Amazons, and Lapiths or Greeks fighting centaurs – symbolizing the struggle between the rational and civilized and the irrational and sensuous. Much of the frieze is now in the British Museum; the Corinthian capital, arguably the earliest known (dating from over 100 years before the birth of Callimachus), has sadly been lost.

The single, central Corinthian column is perhaps the most intriguing detail of the temple. Single, free-standing columns – not part of a structural system – have an ancient history in sacred architecture, marking holy sites or thresholds. The sculptural panel above the Lion Gate at Mycenae, dating from around 1250 BCE, features an image of a single, seemingly free-standing column, and the Biblical description of the temple that Solomon built in Jerusalem, perhaps 3,000 years ago, mentions a pair of columns placed outside the Holy of Holies: 'And he reared up the pillars before the temple, one on the right hand, and the other on the left; and called the name of that on the right hand Jachin, and the name of that on the left Boaz' (II Chronicles 3: 17).

The holy of holies at Bassae – the adyton, or inner sanctuary – lies immediately to the south of the free-standing column, which consequently marks a significant threshold. In this sanctuary stood, no doubt, an image of Apollo that at dawn would have been bathed in sunlight entering through a window or perhaps originally a door placed in the sanctuary's east wall.

The mixing of the three orders in one temple is certainly unusual, but it is not irrational. Vitruvius, when describing the origin of the orders, mentions their sexual attributes – masculine Doric, matronly Ionic, and maidenly Corinthian – so if these associations were traditional, and reflected Greek thinking, then to mix the orders in a temple was merely to fuse the feminine with the masculine. Consequently, to enter the Temple of Apollo Epicurius and to penetrate to its holy of holies is to make the journey from the world of man, through that of woman, to the 'maidenly' Corinthian column – a sort of petrified virgin priestess – and into the presence of Apollo, here venerated in his feminine role as 'the helper'.

This stabilized ruin remains one of the most moving of Grecian temples.

3

THE PANTHEON
Rome, Italy | c.118–128 CE

The Pantheon is one of the most sublime – and most emotive – buildings created in the Roman world, and in many ways is now the best preserved. In its daring scale, its brilliantly engineered and robust construction, and its use of pioneering building technology, it is the epitome of heroic and breathtaking Roman architecture.

The Pantheon is the creative marriage of three primary forms, the immutable and cosmic virtues of which were extolled by Plato. The triangular form of the great portico, with its series of massive monolithic columns from North

Africa, sits in front of a huge drum that rises from a circular plan, while the interior volume of the building, from the ground to the top of its vast domed roof, fits within a cube.

And then there are the materials and astonishing techniques of construction used in this 1,900-year-old masterpiece. All is experimental and bold: the mighty coffered dome, with a span of 43 metres (142 ft), is made of mass concrete that had to be carefully mixed and cast to prevent it cracking and collapsing as it dried, and the huge weight and lateral thrust of the dome are contained and carried to the ground by means of a series of ingenious engineering solutions.

The original function of the Pantheon had a profound influence on its form. It was built between 118 and 128 CE, during the reign of the emperor Hadrian, to replace a temple from Agrippa's time that burned down in 80 CE. The Pantheon was, as its name implies, built to be a temple to all the gods. Since it was a home to all the gods, each god had to have its own home within the temple. This meant the construction of a series of niches – ultimately seven in number, with the main door occupying the place of an eighth niche. These niches, each inhabited by an image of one of the gods, were constructed within the thickness of the wall, as were a series of smaller niches at ground and upper level; they not only gave distinction to the interior but also determined the way the Pantheon was constructed. The drum that supports the dome gains its strength through design, not through mass alone. It is an engineered, honeycomb affair, incorporating relieving arches that transfer loads and give extra strength.

The monolithic granite columns from Egypt were originally placed on a low plinth or stylobate in the manner of earlier Greek temples, while the hemispherical concrete dome was a form pioneered by Roman engineers and one that could not have been built by earlier ages.

Presumably the point being made was that Hadrian's empire united Egyptian, Greek, and Roman culture, and this unity was represented by the Pantheon – a structure that no earlier civilization could have made. It represents Roman technological prowess married to ancient forms, in order to allow a huge internal space to be covered by a dome of unprecedented size that floats above the ground without the need for any support other than the wall on which it sits.

To enter the Pantheon is always overwhelming. This lofty space proclaims the power of primary geometry: the whole of the cylinder and dome would fit within a mighty cube, so the diameter of the round floor is the same as the distance from floor level to the top of the dome, while the distance from the floor to the top of the cylinder is the same as the distance from the top of the cylinder to the top of the dome. It's all so straightforward, so deeply pleasing – and the meaning is clear. In here, the world of the gods is made tangible, represented by circular geometry symbolizing eternity – creation without beginning or end. The dome, of course, is the great wonder of the place. Concrete had been used by Roman architects in the construction of large open-plan buildings for a couple of centuries before the dome of the Pantheon was cast – it wasn't a pioneering material, but it had never been used quite like this. It was made with pozzolana (a type of volcanic earth), pumice, stone chips or gravel, sand, and lime, all mixed up with water. The concrete would harden through chemical action between the pozzolana and the lime, and if hydraulic lime was used the concrete would even harden underwater. Concrete had great potential, for it produced structures that were incredibly strong in compression and homogeneous in nature. This was important when used for arches or domes. Usually the weight of a dome exerts a powerful lateral thrust and so demands complex buttressing, but a homogeneous concrete dome has minimal external thrust, with most of its 5,000-tonne weight transferred directly to the wall below. The weight of the Pantheon's dome is decreased, and its strength increased, by the coffering on its inner surface, which reduces its bulk and creates vertical and horizontal ribs that stiffen its structure.

The open oculus at the top of the dome is the main clue to the original meaning of the building. When the sun shines, a bright circle of sunlight, the sun in miniature, moves perceptibly across the coffers. It marks the passing of time, the cycle of the ages, and confirms the symbolism of the dome as the cosmos, as an image of the celestial canopy of the Roman Empire that embraced the known world.

4

PALAZZO RUCELLAI
Via Della Vigne Nuova, Florence, Italy | c.1446–61 |
Leon Battista Alberti

The Renaissance – the rebirth of the sophisticated culture and technological brilliance of the Greek and Roman worlds – flowered in southern Europe during the fourteenth century. With hindsight, the bright dawn of the Renaissance can be seen to have had a slow and complex emergence from the

Gothic world, in large part due to influences from the Arab and Islamic cultures, which had retained much knowledge from the ancient classical world during the chaotic centuries of Europe's so-called 'Dark Ages'.

The Renaissance first manifested itself in Italy, notably in Florence, and was initially expressed most powerfully through literature, famously in the writings of Dante Alighieri and Francesco Plutarch; in the visual arts, by pioneering painters such as Giotto di Bondone; and, from the early fifteenth century, in architecture, where Filippo Brunelleschi and the slightly younger Leon Battista Alberti led the way.

The Palazzo Rucellai in Florence is one of the earliest – and most archaeologically inspired – examples of monumental Renaissance domestic architecture. It stands at the intersection of narrow streets and a small piazza in the city centre, and its façade is generally accepted as the work of Leon Battista Alberti, with the interior and court almost certainly designed by Bernardo Rosellino. Alberti was a humanist author, poet, and philosopher, and was the first great archaeologically minded architectural theorist of the Renaissance, expressing his thoughts on architecture in his seminal *De Re Aedificatoria* of 1452.

The palazzo looks like Alberti's work and can be interpreted as an expression of his ideas about the meaning and exemplary importance of ancient Roman building. But what is certain is its mid-fifteenth-century date – and its restrained but simmering architectural power. The building has the atmosphere of a chained, brooding beast of great strength and purpose, quiet for now but waiting to burst into action. It is full of ideas, some based on a shrewd study of history, others simply novel. The façade of the palazzo, conceived as part of a street and so very much an example of urban architecture, is now famed as one of the first Renaissance elevations to incorporate tiers of pilasters and bands of entablature in a roughly correct proportional relationship.

As has often been observed, this composition appears – and probably was – inspired by Alberti's contemplation of the stupendous ruins of the Colosseum in Rome, one of the most spell-binding and inspirational Roman structures for Renaissance designers. It was not only the vast scale of the Colosseum and its mysterious methods of construction, incorporating brick, stone, and concrete, that captured the imaginations of Renaissance architects, but also the novelty of its design. It represented a supreme example of the Romans' ability to develop and adapt the language of classical architecture to suit the tastes and requirements of their age. The columns, pilasters, entablatures, and pediments of single-storey Greek temples had been adopted by earlier Roman culture and used in Grecian manner, such as at the Temple of Portunus in Rome and at the Maison Carrée in Nîmes, France, both built during the first century BCE. But by the time of the construction of the Colosseum in Rome in 70–80 CE, the conventional application of the orders was not adequate or appropriate.

The Colosseum was to be a huge, multi-storey public building dignified with stone cladding and the application of the classical language of design. So the orders were used, but with columns and entablatures stacked one upon the other. They were no longer a rational expression of construction – columns were not primarily load-bearing and entablatures were no longer at eaves level representing lintels and beamends. Instead they were conceived as part of a scheme of wall decoration comprising tiers of columns – called 'engaged' because they are in fact physically part of the wall – and horizontal bands of entablatures. Together these columns and entablatures define arcades that were conceived to serve as vast windows, allowing the interior of the Colosseum to be lit and ventilated. An element of classical propriety, certainly of hierarchy, was given to this radical reinterpretation of the orders and to their unconventional use as mere ornament, as they were stacked in a specific sequence. The order of columns used at ground level was the masculine Doric, the stoutest of the three Greek canonical orders and the strongest-looking. Next, at first-floor level, the Ionic order was used, lighter and more elongated. Then, at second-floor level, there was the Corinthian – the feminine, the most ornamented and most elongated of the orders. Finally, on the top stage, like the attic on a triumphal arch, the architects placed a tall screen wall embellished with pier-like pilasters on plinths.

This adaptation of the conventional classical composition was inspired, and meant that Greek classicism – with all its ancient cultural pedigree – could continue to dignify ever-bigger, more novel, and more demanding Roman building types.

It seems that Alberti recognized the critically important contribution the Colosseum had made to Roman architecture, seeing it as proof that the ancient language of classicism did in fact evolve – and could again, to suit the special demands of fifteenth-century Italian city-states. So a starting point for his design of the Rucellai façade was the tiered elevation of the Colosseum, with its arched opening providing the model for arched windows.

But Alberti took an inventive rather than doggedly imitative approach when emulating the Colosseum's use of the orders, while accepting the basic hierarchy. For the ground floor he used the Tuscan order, with a permutation of the Ionic above and a simplified Corinthian on the second floor. The façade is topped with a heavy cornice and at ground level a stone bench is built into the design, giving the building a generous role in the public life of the street.

Additional Roman references abound. Alberti faced the wall with rusticated stonework – a favoured Roman treatment for robust structures such as fortifications, in which each stone block is individually expressed – which contrasts with the smooth stone of the entablatures and pilasters. Even the construction of the palazzo makes reference to Roman prototypes because Alberti incorporated Roman-style *opus reticulatum* (reticulated work) formed with diamond-shaped bricks, based on his archaeological explorations.

Other details on the elevation relate to the palazzo's client and its complex and prolonged building history. The client was the wealthy and powerful Giovanni Rucellai, a wool merchant who held political office under Cosimo and Lorenzo de' Medici. Rucellai was a leading creative force in the making of his palazzo. He was something of a classicist and scholar, and one of the city's leading patrons of art and architecture. He financed the completion of the marble façade of Santa Maria Novella – the upper portion of which was also designed by Alberti in 1458 – and he commissioned works by leading artists, including Filippo Lippi and Paolo Uccello.

The palazzo was constructed in phases as and when sites were acquired, with a key moment being the betrothal in 1461 between Rucellai's son Bernardo and Nannina, the daughter of Piero di Cosimo de' Medici. This seems to have stimulated work on the palazzo, and explains the union of Rucellai and Medici devices in the frieze and the use of Medici interlocking rings in the spandrels of the windows. The engagement also prompted the construction of the triumphal-arch-like loggia in the piazza opposite the palazzo, which was completed by the time of the marriage in 1466 and might also have been designed by Alberti.

The ground-floor elevation of the palazzo is lit only by small windows, signifying its utilitarian function. The main entertaining rooms are located on the first-floor *piano nobile*, the main family apartments are on the second floor, and a third floor, for servants and service uses, is concealed from the street.

5

VILLA ALMERICO CAPRA (LA ROTONDA)
Via Della Rotonda, Vicenza, Italy | 1566–71 | Andrea Palladio

Andrea Palladio, born just over 500 years ago in northern Italy, was the leading architect of the Italian late Renaissance and became – through his work and his book *I Quattro Libri dell'Architettura*, published in 1570 – one of the most influential architects of all time.

The perennial power of Palladio's architecture, notably his assured use of a system of related proportions, suggests that he possessed a 'key' to ultimate architectural beauty – but the nature of this beauty, and the means by which it can be achieved, was not proclaimed by Palladio in direct manner. Instead it has to be deduced from his buildings, and by an interpretation of his writings and those of some of his contemporaries. And one of the best buildings to help crack the code of Palladio's creative genius – and his inspired use of history – is the Villa Almerico Capra, otherwise known as the Rotunda.

The villa was built quickly between 1566 and 1571 and although not completed to his published design, and with most interior decorations being of later date, it is regarded by many as the quintessence of Palladio's classicism. The building is dominated by the details and forms that are characteristic of Palladio's architecture. It is absolutely symmetrical and inspired by the visual and geometric power of the sphere and the cube, and by the use of harmonically related proportions. Its beauty comes not from ostentatious ornament but from bold simplicity – from a unity that combines all elements of the composition, and from the rational expression of construction.

Palladio gives little away in his book when discussing the design. He extols the beauty of the site – 'it is encompassed with most pleasant risings, which look like a very great theatre [with] most beautiful views' – and describes the house in simple terms: 'there are loggias made in all the four fronts ... the hall is in the middle, is round, and receives its light from above.'⁴

To understand more about Palladio's architectural thinking and the design of this house, it is necessary to delve deeper.

In his book Palladio included, along with reconstructions of exemplary Roman buildings set in flattering proximity to his own designs, a description of the seven proportions that he declared would produce the 'most beautiful' rooms. These are a circle, a square, a square and a third, the diagonal of a square, a square and a half, a square and two-thirds, and a double square.

The key thing is that all these proportions are simple extensions of, or are closely related to, the same basic unit – the square, or its three-dimensional equivalent, the cube. Even the diagonal of a square proportion, if extended, becomes a double square. So this proportional system, organized around the square, creates a series of harmoniously related shapes.

For Palladio, this integration – the fact that the proportions used were 'commensurate' – was vitally important; it was the key to getting all the elements of a building – its plan, its elevations, details such as window and door openings – to relate in a pleasing manner. If this harmony were not achieved, then beauty would be absent. But, although Palladio never admits as much, it is clear that the application of these related proportions was far more than just a way of achieving a harmonious design.

Palladio, along with many during the Renaissance, believed that proportion was divine and represented the very building blocks of God's creation. For Palladio, his proportional system enshrined immutable laws of nature that were first observed, codified, and applied by the Greeks and Romans and which lay embedded in their buildings. To design according to the architectural principles of classicism and to emulate its proportional systems was to echo the beauty of nature itself.

The cube and square dominate the design of the Villa Almerico Capra: the cylindrical and domed central hall was intended to occupy a double cube volume, a number of rooms are a square and a half in plan, and window openings are permutations of a square. The entire building is proportionately connected and resonates like a piece of carefully composed music.

For Renaissance thinkers, visual and musical harmony were the same. Palladio made this view clear in 1567 when he wrote that 'just as the proportions of voices are harmonious to the ears, so those of measurement are harmonious to the eyes', while one of his patrons, Daniele Barbaro – a high-born Venetian diplomat, philosopher, and cleric with whom he collaborated on a translation of Vitruvius – explained in 1556 that 'beauty results only from the right proportions' and that 'numbers have a divine power when the proportions are consonant ... [Singers'] voices should be in tune, and the same applies to the parts in architecture. This beautiful manner in music as well as in architecture is called harmony, mother of grace and delight.' Proportion, emphasized Barbaro, is 'the glory of architecture ... the miracle [that] contains all the secrets of the art'.

The Villa Almerico Capra suggests very strongly that these bold assertions are true.

6

CHÂTEAU DE VAUX-LE-VICOMTE
Maincy, Seine-et-Marne, France | 1658–61 | Louis Le Vau, Charles Le Brun, André le Notre

The Château de Vaux-le-Vicomte is among the most influential and most beautiful country houses ever built. It was also, for its creator, an act of tragic folly.

The château was commissioned by Nicolas Fouquet, who was the superintendent of finances for King Louis XIV and one of the richest and most powerful men in France. The new building was to express these latter attributes, as well as Fouquet's exquisite and up-to-the-minute artistic taste, on a princely scale. He chose as his architect the emerging genius Louis Le Vau – a royal favourite who, in 1659, collaborated on the design of the Palais du Louvre in Paris and, in 1661, started to extend the Palais de Versailles. Fouquet selected the royal gardener André le Notre to create the gardens around his château, and Charles Le Brun – the king's favourite painter – to oversee the interior embellishment. Fouquet was determined to create a dazzling masterpiece that even a king would envy.

The house and park this group of men produced was indeed a masterpiece, and one that set in motion architectural ideas that rippled through the northern countries of Europe. It is hardly surprising that Vaux-le-Vicomte anticipated Versailles in some notable ways – after all, the same designers were involved – but it is nevertheless remarkable, considering that one was only a château for a courtier while the other was a palace for a king who, in the 1660s, could reasonably claim to be the international leader of taste and the most powerful monarch in the world.

The most obvious harbinger of Versailles' glories is the garden at Vaux-le-Vicomte. By 1661 Le Notre was working on the gardens at Versailles to make them the largest and most magnificent in the world, but it seems he used his time at Vaux-le-Vicomte to evolve some of the key ideas that he was to develop later at Versailles. Both involved the controlled manipulation of nature and landscape to create formal gardens, largely symmetrical, with lavish water features combined with geometric parterres and avenues, all organized around a powerful central axis that the house sits astride. In addition, at Vaux-le-Vicomte Le Notre indulged in much visual wit, distorting features in a secretive manner (termed *anamorphosis abscondita* at the time) to create false perspectives and confuse scale in order to make the garden, when viewed from the house, appear longer and larger than it is.

Architecturally the design of the château is dependent on the early seventeenth-century Baroque of Rome, but with a touch of Palladio, as revealed by the strict symmetry of its plan – the way in which the house is conceived in the round, and the domed garden front that can be seen as a reference to both Palladio's Villa Almerico Capra and the Pantheon. Also Palladian is the centrally placed entrance vestibule, with its square plan form and pair of flanking staircases, set on axis with a double-height salon.

But the movement of the elevations owes much to the Italian Baroque, as does the oval form of the salon, perhaps derived from the salon in the Palazzo Barberini in Rome, built between 1627 and 1633 to the designs of Gian Lorenzo Bernini and Francesco Borromini. The external expression of this salon as a curving, centrally placed bay captured the imagination of contemporary architects and became one of the enduring features in country house design for the next 200 years. Another influential aspect of the design is the Ionic pilasters, rising from ground level through two storeys to the eaves cornice and known as a 'giant order'. This was used to ornament the château's elevations.

Other aspects of the château that were novel, and that had a profound influence, include its 'double-pile' plan (usually châteaux at this time were one room deep, whereas Vaux-le-Vicomte is two) and corridors at basement and first-floor level to increase privacy and convenience.

Early and notable expressions of the influence of Vaux-le-Vicomte in the British Isles include Easton Neston, Northamptonshire, and Chatsworth House, Derbyshire. The latter, started in the 1680s, sports a giant order and has a north front of c.1705 with a full-height curved bay. American examples include the White House, of 1792, with a full-height curved bay in the centre of its south façade.

The interior of Vaux-le-Vicomte is magnificent – in particular the salon, which is a sublime architectural space, with its arcaded walls treated as exterior elevations and its domed ceiling painted as a cloud-strewn sky. On each side of the entrance and salon at ground floor is an apartment; Fouquet fitted one of these up in grand style for the king, to encourage him to be a frequent visitor.

Upon completion of the château in August 1661, Fouquet held a lavish and celebratory fête with the king as guest of honour. The consequences could not have been more dreadful. Upon seeing the princely house and Fouquet's lavish generosity to his guests, the absolutist monarch, his mind already poisoned by Fouquet's jealous rivals, felt upstaged and concluded that such a splendid creation could only have been funded by embezzling royal revenues. Fouquet's elimination was decided and, after skilful plotting to isolate him from power, he was arrested three weeks later, endured a lengthy and unprincipled trial, and in 1664 was imprisoned for life with his wife exiled. Meanwhile, the château had been sequestered by the crown, and the king, determined to outdo this upstart, launched the project that led to the creation of Versailles. As

Voltaire later observed, on the evening of the fête Fouquet was like a king in his palace, but by two in the morning he was a nobody. Rarely can a fall from so great a height have been so rapid and absolute – and mostly due to architecture.

7

MARSHALL, BENYON, AND BAGE'S FLAX MILL

Ditherington, Shrewsbury, England | 1796 | Charles Bage

This flax mill is the oldest iron-framed building in the world and is now popularly, and with good reason, regarded as the father of the skyscraper. It introduced metal-frame construction technology that, from the late nineteenth-century when combined with other developments such as the mechanically operated elevator, made high-rise architecture possible and practicable.

The mill does not look like a pioneering metal-framed skyscraper. It is only five storeys high and its outer walls are of red brick, with the only really notable feature being the plain and utilitarian nature of the elevations, the close spacing of the large windows (many now blocked or reduced in size) designed to allow the interior and its working areas to be flooded with light. Also remarkable is the unusually large size of the bricks. This reveals something of the economic climate in which the mill was constructed and within which its owners operated. In 1784 the British government introduced a tax on newly made brick in order to recoup some of the vast sums lost fighting the abortive war with the American colonies. Bricks were initially taxed at the rate of 4 shillings per thousand. But brick makers and builders soon came up with a dodge. If individual bricks were made to a larger dimension, the same tax would be paid, but fewer would be needed per construction project. So after 1784 some brick-makers started producing bricks as large as 11 × 5 × 3.5 inches (280 × 125 × 90 mm), in contrast with the traditional 8.5 × 4 × 2.5 inches (215 × 100 × 65 mm). In response, the government plugged this loophole by introducing tax penalties on bricks with a volume greater than 150 cubic inches (2,500 cu mm). The tax was increased in 1794, 1797, and 1805, peaking at 5 shillings and 10 pence per thousand. The tax was not repealed until 1850. The bricks used at the Ditherington Flax Mill measure 10 × 4.5 × 3 inches (250 × 115 × 75 mm) – so they are larger than usual, but just below the 150 cubic inch maximum.

It is the interior that makes the mill of international importance. The Industrial Revolution had gradually made cast iron a familiar building material, but usually just for columns designed to carry compressive loads and incorporated with timber elements – notably beams – that had far more tensile strength than cast iron. A sensational and early construction entirely of cast iron is the Iron Bridge, started in 1777 at Coalbrookdale, Shropshire, and created by architect Thomas Farnolls Pritchard and ironmaster Abraham Darby III.

But at Ditherington a coherent structural system was devised with cruciform-section stanchions of cast iron supporting cast-iron beams to form a multi-storey structural frame. Brick was used for the outer load-bearing walls and to form shallow, curved jack-arches, that spring between beams, to serve as ceilings and floors. The roof structure, which gives the mill a characterful gabled silhouette, is also made of cast iron.

This structural system was strong and capable of supporting heavy machinery; it was also relatively quick and cheap to construct, with the main cast-iron structural components being mass produced in factory conditions. It was assumed at the time – wrongly, as it turned out – that cast iron was safe from fire damage and that this mill was essentially a fireproof structure. In fact, cast iron, which is brittle by nature, does not perform well in fire and readily cracks when doused with water, as was revealed in Ditherington in 1811 when part of the mill caught fire.

The architect for the mill was Charles Bage, who seems to have taken some inspiration from the Derby-based civil engineer William Strutt, then an acknowledged expert on, and the leading designer of, fire-resistant textile mills, using cast iron and other materials deemed to be fire-resistant at the time.

The mill is a powerful and evocative structure – and now also intensely melancholic, for after decades of near dereliction it continues to stand gaunt, with its future uncertain.

8

DULWICH PICTURE GALLERY
London, England | 1811–17 | Sir John Soane

John Soane was one of Britain's best and most fascinating architects. He was born in 1753, the son of a bricklayer, and through family connections entered the office of George Dance the Younger to train as an architect at the age of 15. Dance, architect and surveyor to the City of London and a brilliant and inventive designer, offered Soane not just sound advice and inspiration but also useful connections and opportunities of patronage. Dance was a founding member of the Royal Academy and encouraged Soane to join its schools in October 1771. This gave Soane access to a first-class library and to lectures on architecture and drawing. While at the Academy, Soane won a Gold Medal in 1776 for his design of a Triumphal Bridge, and in 1777 he began a travelling scholarship that, in March 1778, allowed him to undertake a modest 'Grand Tour' via France to Rome.

Soane was immensely talented and immensely ambitious. Also – fortunately – he was energetic, conscientious, and willing to work hard. He also knew that gaining knowledge was only part of the challenge. He trained himself to observe and to analyse, in order to discover the essence of architectural beauty and how to achieve it. More than anything else, this determination to see with fresh eyes is the secret of Soane's later success as an architect. He wanted to build on first principles and achieve architectural power and dignity with the minimum of means. Ultimately this led to a very individual architecture that, although rooted in history and the classical tradition, was novel and even radical. Soane sought to reinvigorate classical design, and this he achieved through reduction, through abstraction, and through the use of primary, almost elemental forms. The resulting architectural minimalism – combined with an almost obsessional use of natural light (both to enliven his often layered elevations and to illuminate his interiors by means of roof lanterns), and his quest to 'liberate' interior space by creating interconnected rooms – makes much of Soane's work feel intensely modern.

These ideas are expressed forcefully within the buildings he designed between 1792 and 1824 at 13–14 Lincoln's Inn

Fields, which were to serve as his home, private museum, picture gallery, and office. But arguably Soane's most characteristic and pioneering single building – certainly one that well represents his architectural principles and approach – is the Dulwich Picture Gallery. It is pioneering not just because of its sophisticated simplicity but also because it is the world's first free-standing picture gallery open to the public – in a sense the world's first public museum.

The evolution and nature of the building are curious and complex. Dulwich College, founded by Edward Alleyn in 1619, received a bequest in January 1811 of 360 paintings – most of the highest quality and including works by Rembrandt, Rubens, and Van Dyck – from Sir Peter Francis Bourgeois and Noel Desenfans. The bequest included £3,000 for works to the existing College buildings and chapel to house the paintings and the bodies of the two benefactors and, at some later date, that of Mrs Desenfans. The bequest also 'recommended' that Soane, who had been Bourgeois' friend, be commissioned to undertake the works.

This set in train an extraordinary series of events. Soane first visited the site on 8 January 1811 – the day after Bourgeois' death – and thus revealed his almost morbid fascination for the project. Soane's architectural vocabulary had long included the motifs of death, mostly gleaned from Roman tombs, and the fact that the project was to incorporate a mausoleum clearly fascinated him.

Soane rapidly produced a series of designs – combining mausoleum, gallery, and even almshouses – that explored different plans and in which the detailing was gradually pared back. By May 1811 the essential idea for the galleries, top-lit by lanterns and linked visually and spatially by simple arched openings aligned to create a long vista, had been arrived at. These designs, with their generally windowless walls (good for security and to increase wall area for picture hanging), roof lanterns supported on shallow pendentive domes, and almost abstract incised mouldings of classical derivation, related to halls that Soane had been executing since the early 1790s for the Bank of England.

By mid-May 1811 Soane had produced five schemes – all of which he or the College found unsatisfactory. He then produced three more and in July 1811 a design was finally accepted. But even this design was to evolve; indeed, it seems to have evolved during the meeting where it was presented, because one of the elevations shows the arcade – which was eventually constructed – being added sketchily in pencil.

The final design, which got under way in October 1811, was dominated visually and architecturally by the mausoleum, which, standing in the centre and forward of the west elevation (the traditional orientation of death in the ancient world), is cubical in form, crowned by a tall lantern, and has idiosyncratic pilasters and entablature, which support huge stone-made sarcophagi. Inside there is a circle of Doric columns supporting a shallow dome.

Flanking the mausoleum were almshouses (incorporated into the gallery in 1880) and parallel to these Soane placed five top-lit, Pompeian red-painted gallery rooms, interconnected by tall and aligned semi-circular arched openings. The external elevations are executed in brick and articulated and embellished not with the language of conventional classicism but with pilaster strips, blank and open arcades, and horizontal slabs of stone simulating cornices. All is evidently intended to evoke the elemental, almost primitive essence and origin of classicism.

The gallery was opened in 1817 and despite being extended, altered, and repaired after bomb damage, it remains one of the most innovative and inspirational buildings in Britain.

9

THE PALM HOUSE
Kew Gardens, London, England | 1844–8 |
Decimus Burton, Richard Turner

The Palm House at Kew is one of the most significant structures created in nineteenth-century Britain. It combines the precise utilitarian demand of its function with a sense of form to create one of the most beautiful buildings of its age, but more importantly, it is the first building in the world to incorporate the large-scale and structural use of wrought iron. It was a true pioneer and harbinger of the structural world to come, in which wrought iron, because of its superior tensile strength, replaced cast iron as the material with which to realize the massive, wide-span building types demanded by the age. These ranged from huge station sheds such as that built at St Pancras in 1867 to high, wide-span mega-bridges like I. K. Brunel's stupendous Royal Albert Bridge at Saltash on the Devon–Cornwall border, completed in 1859, which utilized vast, tubular 'lenticular' trusses made of wrought iron to carry the rail track across a wide river crossing. And it all started with a palm house in Kew Gardens.

In fact, greenhouses play an important role in the development of engineered construction in the nineteenth century. Joseph Paxton was appointed superintendent of the gardens at Chatsworth House, Derbyshire, in 1826, and in 1836 designed the epoch-making 'Great Stove' conservatory, which was the largest greenhouse in Europe. It was also a showpiece of innovation. It utilized the newly developed ridge-and-furrow glazed roofing system, supported by arched laminated-timber ribs and cast-iron columns, which doubled up as rainwater downpipes. Decimus Burton had been consulted over the design, but Paxton was recognized as the force behind it, and the fame and high reputation of the building (tragically demolished in 1920), as well as that of his slightly later Lily House at Chatsworth, were instrumental in the choice of Paxton to design the vast, cathedral-like hall to house Prince Albert's 1851 Great Exhibition in Hyde Park.

Although it was a temporary structure, and because of its utilitarian form was regarded by many contemporaries as little more than an overblown shed, the Great Exhibition building was ultimately to change the nature of modern architecture. It incorporated more iron than the Chatsworth structures, largely in the form of mass-produced cast-iron components that allowed the building to be constructed to a large size at remarkable speed. Paxton also capitalized on new glass-making technology, perfected in 1848, for the manufacture of large sheets of cast plate glass, and used this new material so effectively that his exhibition hall was christened the Crystal Palace. Also, in the spirit of the inquiring age, Paxton took inspiration from natural history, and studied in particular the structural systems of large flowers and leaves, such as the huge Victoria amazonica water lily, recently discovered by European botanists. He later admitted that its huge floating leaves, with their strong construction achieved via radiating ribs connected by cross-ribs arranged in a herringbone pattern, inspired the design of the roofs at the Crystal Palace.

The Palm House at Kew was constructed at a crucial moment between these two famed Paxton creations and was, in its pioneering and large-scale use of structural wrought iron, arguably ahead of both. In many ways it was the most relevant prototype for cutting-edge engineered construction.

Although Decimus Burton is credited with the design of the Palm House, it is essentially the inspired creation of iron-master Richard Turner, who, with architect Charles Lanyon, had constructed the superb cast iron and glass palm house in Belfast's Botanic Gardens in the late 1830s.

The object of the Palm House was to further Kew's ambitions in the development of 'economic botany' – essentially the procurement, study, nurture, and dissemination around the empire of economically valuable plants such as rubber trees and cinchona, which renders quinine. To achieve its aim of providing a congenial home for tropical plants, the Palm House had to be capable of generating and sustaining the hot and humid atmosphere of a rainforest and accommodating some plants of great height. This Turner achieved by the maximum use of glass, heating systems, and tall and dramatically curving vaults. And it was with these vaults that Turner pioneered the architectural use of wrought-iron ribs, whose tensile strength allowed them to carry substantial lateral loads, combined with cast-iron columns that are massively strong under compression. It is probable that, as an early example of 'technology transfer', this vaulted structural system was inspired by the fabrication techniques developed in the construction of iron-hulled ships.

The Palm House is generally utterly utilitarian, and visually a direct expression of its materials and means of construction, and of its function. As an essentially practical building, its designers appear to have been liberated from the need to dignify it with the application of ornament or to associate it with an eminent historic style. Yet some of the cast-iron components do possess discreet ornaments derived from history and so presumably are, in the spirit of John Ruskin, a modest attempt to elevate pure engineered building into culturally aware architecture.

This marriage between utilitarian construction and architecture is echoed by Henri Labrouste's contemporary and sensational Bibliothèque Sainte-Geneviève in Paris. Completed in 1850, it combines boldly expressed iron construction with a rich repertoire of applied decoration and forms – such as domes and a basilica plan – that evoke powerful historic precedent.

10

E. V. HAUGHWOUT BUILDING
488–92 Broadway, New York, USA | 1857 | John P. Gaynor

The development of cast-iron-fronted architecture evolved in parallel with the development of iron-framed architecture. These two types of construction, both of which utilized modern building technology, overlapped in certain significant ways, although iron-fronted building evolved a distinct character, largely due to the nature of cast iron. The fashion for cast-iron-fronted buildings, which realized the decorative as well as the structural potential of the material, emerged in the 1850s and continued into the 1870s. Outstanding examples survive in the SoHo area of New York (notably in Greene Street), and in Glasgow, which incorporate cast-iron components, usually of classical form, assembled to make entire street frontages. This system allowed ornate façades to be constructed quickly and relatively cheaply. They had huge load-bearing capacity and were believed to be relatively fireproof. The first iron-fronted building in New York – and probably the world – was erected on the corner of Duane Street and Center Street in 1848–49 by James Bogardus, an inventor fascinated by the possibilities offered by iron for prefabricated architecture using mass-produced components. This was an industry subsequently developed by Glasgow ironmasters, who exported cast-iron building components around the world.

One of the earliest cast-iron-fronted buildings in the British Isles is Gardner's Warehouse, in Jamaica Street, Glasgow. It was built in 1856 to the designs of John Baird and looks astonishingly modern with its arcaded and classically composed but minimally detailed four-storey elevation, sparkling with large sheets of plate glass (an innovation of the 1840s). The Ca' d'Oro Building in Gordon Street, Glasgow, is another fine if more clearly culturally attuned example of the type, with a design closely based on historical precedent. Designed in 1874 as a warehouse by John Honeyman, it has a cast-iron façade inspired by fashionable Venetian Gothic architecture, which consciously proclaims the warehouse to be a piece of architecture rather than a mere utilitarian building – though perversely, the detail is classical rather than Gothic. The interior was also largely iron-framed, with a system of masonry-arched floors. All was intended to be fireproof, but the building was severely damaged by fire in 1987, when most of the interior was destroyed. During reconstruction, the glazing in the iron façades was restored to its startling original and simple form, with large sheets of plate glass set within the iron frame.

Interestingly, the avant-garde technology of the façade was often not continued internally. Many of New York's iron-fronted buildings of the mid-nineteenth century are structurally fairly traditional, with brick party walls buttressing the façades, and, internally, timber beams and joists and even sometimes timber posts used as well as cast-iron stanchions. Usually additional fire protection was achieved by cladding ceilings with metal sheeting, embossed with ornamental patterns.

One of the earliest and best New York examples is 488–92 Broadway. It was constructed in 1857 to the design of John P Gaynor, with the cast-iron façade – of splendid Renaissance form, with arched windows set, Colosseum-fashion, between tiers of columns and entablatures – made by the Daniel D Badger ironworks. Occupying a corner site, the building has two cast-iron façades, and unlike most other New York examples, these façades are not just ornamental but form the building's primary structure and are connected by an internal frame integrating timber beams and joists with iron girders and columns. This combination of load-bearing iron façade and partly iron-framed interior makes this building a key and early – if still exploratory – example of the technology that was to underpin the construction of the city's high-rise buildings later in the century.

This building contains another component that is even more important in the development of the high-rise. It is only five storeys high, but its owner – the high-class Haughwout store, which sold expensive china, cut glass, and silverware – did not want its illustrious customers, who included Mrs Lincoln on shopping trips to furnish the White House, toiling up and down stairs to the showrooms. So in March 1857, a steam-powered hydraulic 'safety elevator' was installed; made by the Otis Company, this was the first passenger elevator in any building in the world. The appearance and function of the 'ascending room' was described at the time by the *New York Tribune*: 'Among the novelties of the building is an elevator to be worked by steam … furnished with a sofa and carriage to carry ladies from one floor to another.'

People flocked to the store to see this new contraption; invented by Otis in 1852 and demonstrated in 1853 at New York's Exhibition of the Industry of All Nations, it was not only one of the wonders of the age but was to revolutionize architecture by making high-rise living possible.

At about the same time that the Otis elevator was installed in the Haughwout Building, the proprietors of the evidently go-ahead Gardner's Warehouse in Glasgow also acquired one, making it probably the first elevator to be used in Britain.

11

ORIEL CHAMBERS
Water Street, Liverpool, England | 1864 | Peter Ellis

A key building in the nineteenth-century development of iron-fronted and iron-framed construction, utilizing large areas of plate glass, is Oriel Chambers in Water Street, Liverpool. In many ways it's a maverick and rather oddball piece of construction. Designed by Peter Ellis and completed in 1864, the building – conceived as self-contained sets of offices or chambers – is structurally advanced, although, being clad in stone, this is not immediately apparent. The primary structure is a cast-iron frame, formed with H-section stanchions supporting inverted T-section beams set at narrow spans that do not require great tensile strength. Floors are formed by masonry 'jack arches' turning between beams.

The iron frame of the building's street elevation is concealed behind stone cladding, embellished with simple historically inspired details, which now gives the building a more traditional appearance. When it was first completed, however, the building's most obvious novel feature – the large plate-glass bays or oriels that project between the slender stone-clad piers – was enough to shock, puzzle, and offend conventional-minded observers.

As is often the fate of avant-garde artists, Ellis was too far ahead of the tastes of his time; he was misunderstood, and his intentions brutally misrepresented. The influential journal *The Builder* described the building, soon after its completion, as 'a large agglomeration of protruding plate-glass bubbles', sneering that 'did we not see this vast abortion – which would be depressing were it not ludicrous – with our own eyes, we should have doubted the possibility of its existence. Where and in what are their beauties supposed to lie?'[5] This question in now easily answered. The 'beauties' lie in the novel, inspired, and appropriate use of new construction technology to create a building that is superbly fitted for its site and function.

Nowadays, Oriel Chambers is famed for arguably introducing the notion of the 'curtain wall' that became one of the hallmarks of twentieth-century Modernism. In essence, a curtain wall is a lightweight, generously glazed screen – typically with large panes of glass set in a minimal structural frame – that has no load-bearing responsibility beyond supporting its own weight. This, of course, means that structural elements play a minimal role, allowing glazing to predominate. The street elevations of Oriel Chambers are obviously not curtain walls in the twentieth-century sense; although the glazed oriels are large, allowing much light to flood inside, they are not part of a continuous system of glazing but are still separated, in conventional manner, by stone-clad iron stanchions, and are essentially no more advanced than the elevations of Gardner's Warehouse in Glasgow. The real wonder of the building is hidden in its courtyard, and seemingly not even seen by *The Builder*'s critic. Here there is a true curtain wall – an all-glass façade, liberated from a structural role, with panes set in timber frames of slender form, and with each storey stepping back slightly to allow maximum light to reach the more shaded lower floors. This is a wonderful sight: the triumph of unadorned functionalism marking the arrival in 1864, in an obscure Liverpool courtyard, of the world of architectural Modernism. In 1969, in the South Lancashire volume of his monumental forty-six-volume survey *The Buildings of England*, the architectural scholar Nikolaus Pevsner described Oriel Chambers as 'one of the most remarkable buildings of its date in Europe'.[6] In 1936, in his *Pioneers of the Modern Movement*, Pevsner had praised the rear curtain wall as 'almost unbelievably ahead of its time'.

This was not a one-off for Ellis. In 1866, at 16 Cook Street, Liverpool – another office development – he designed a street façade with slender and simple stone mullions, large areas of glazing, and a rear elevation that once again has a curtain wall, but that also incorporates an astonishing and beautiful fully glazed cylindrical staircase tower that is incredibly pioneering in its elegant and functional minimalism and utter absence of ornament. Such ruthlessly utilitarian details might possibly have been found on greenhouses or station buildings at the time, but were unprecedented for urban, commercial architecture.

The international influence of this pair of buildings was, in all likelihood, direct and dramatic. In August 1864 John Wellborn Root was sent to Liverpool by his father to escape the consequences of the South's looming defeat in the American Civil War. Root, with an interest in architecture, must have studied the novel Oriel Chambers because its structural system and details had a clear influence not only on Root's own work but on early modern architecture in the United States. At the end of the war Root left Liverpool for New York, where he studied architecture, and by 1871 he was established in Chicago, where he soon went into partnership with Daniel Burnham. Together they helped forge what is known as the Chicago School, characterized by metal frame construction, masonry (often terracotta) façades, large areas of plate glass, and projecting bay windows. Oriel Chambers, where in essence all these elements are present, can be seen as the direct inspiration for Root and Burnham's trailblazing Chicago projects, including the Rookery of 1887 and the epoch-making highrise Reliance Building of 1890, completed in 1895 to the designs of Charles Atwood.

12

ST PANCRAS STATION (INC. THE TRAIN SHED)
London | 1865–68 | Henry Barlow
Midland Grand Hotel | 1865–76 | George Gilbert Scott

Mid-nineteenth-century Britain was the setting for the emergence and ultimate expression of two parallel but dependent visions of how to build for the modern world. On the one hand there was the quest for architecture, dignified by reference to admired historic precedent, that was believed to be an appropriate expression of the power and glory of Britain and its empire – the largest, wealthiest, and most powerful the world had ever seen. On the other hand there was the dramatic development of engineered construction that, rooted in the stupendous technological innovations of the eighteenth-century Industrial Revolution, allowed for the realization of new building types of unprecedented scale and potential, wrought in pioneering and utilitarian manner, utilizing novel materials and means of construction.

These two parallel worlds – ornamented, history-inspired architecture and practical, unadorned construction – had been defined in memorable manner in 1849 by influential art critic and theorist John Ruskin in *The Seven Lamps of Architecture*. For Ruskin, 'architecture proposes an effect on the human mind, not merely a service to the human frame', and is the 'art which ... adorns the edifices raised by man' and makes a building 'agreeable to the eye ... or honourable by the addition of certain useless characters'. Ruskin clarified his view by taking the example of functional fortifications: 'no one would call the laws architectural which determine the height of a breastwork or the position of a bastion. But if to the stone facing of that bastion be added an unnecessary feature, as a cable moulding, that is Architecture.' For Ruskin it was the addition of the practically 'useless' but beautiful ornament – perhaps rich in historic association or artistic skill – that raised mere building to the status of architecture. And it was these two worlds that met in epic and mutually beneficial manner in the project to create a new railway terminus and related hotel and railway offices at St Pancras for the Midland Railway Company.

The St Pancras story started in June 1863, when the Midland Railway Company obtained an Act of Parliament to extend its line into London and to construct a terminus at St Pancras. The terminus was designed by Henry Barlow, and its main feature, started in November 1867, was the stunning iron-built train shed, which in scale and engineering elegance was, when completed, one of the wonders of the age. It achieved a span of 75 metres (245 ft), making it the widest single-span structure in the world, and was executed in a boldly utilitarian manner wherein the prime ornament was the functional beauty of its construction.

The engineering principles of this giant shed are elegant and to the point. The huge wrought-iron ribs meet at a pointed apex – not to evoke the Gothic spirit, but because Barlow took the view that this design performs best to resist wind pressure. The ends of opposing ribs are linked at ground level by wrought-iron ties to help restrain the arched roof from spreading and to form the base on which the train tracks and platforms stand. All the building's key elements and forms are integrated and ultimately determined by functional requirements.

* * *

The Midland Grand Hotel at St Pancras is one of the finest expressions of the nineteenth-century Gothic Revival in architecture. Until the early nineteenth century in Britain, classical design – gradually ushered into the country from the early sixteenth century – had become the vernacular of British building, but soon after 1800 other forms of architectural inspiration were sought and followed. The Rococo fashion of the mid-eighteenth century had popularized a playful and historically incorrect revival of medieval Gothic, along with Chinoiserie, Saracenic, and Hindoo. And in the later eighteenth century new archaeological discoveries and explorations of the antique past reinvigorated classical design, leading to neo-classicism and the Greek Revival. Tastes became increasingly eclectic, even embracing versions of Egyptian architecture inspired by a wave of interest following Napoleon Bonaparte's adventures around the Nile. All history, it seemed, could be pillaged in the quest to find a modern architecture that was novel, decorative, noble by association with past greatness, and titillating. No one questioned the basic concept that a modern architecture need to refer to the past, and no one argued that decoration for its own sake, that was not the consequence of a structural system, was irrelevant. Even imaginative and artistically radical architects like John Soane did not utterly reject history and ornament; they simply wanted to reinvent, minimalize, and – to a degree – abstract. But in the 1830s things changed in Britain and architecture moved in a direction that was unprecedented and uncompromising. Architectural design became a moral issue. There was, it was argued, ethically and objectively a right and a wrong way to build – and in consequence, architecture would change forever.

In the early decades of the nineteenth century, the notion emerged that Britain should develop a dignified national style that reflected its distinct history, that was appropriate for the diverse and demanding modern age, and that reflected the nation's Christian tradition and imperial power. The opportunity to express this conviction in a tangible manner was presented in October 1834 by the catastrophic destruction by fire of the Palace of Westminster, the home of the British Parliament. The competition brief for the rebuilding of the palace specified that the design was to be in the Gothic or Elizabethan style because these, it was argued, were 'British', represented national identity, and would allow the new building to sit in harmony alongside the Gothic Westminster Abbey. The competition was won by Charles Barry, whose Perpendicular Gothic design had been detailed by the fanatical Gothic Revivalist A W N Pugin. Work started in 1840, by which time Pugin had started to produce polemical texts that, accompanied by his designs, transformed British architecture.

Pugin, a convert to Roman Catholicism, asserted in his writings that Gothic architecture – Christian in origin – was spiritually, artistically, structurally, and ethically superior to 'pagan' classicism. In his key text, *The True Principles of Pointed or Christian Architecture*, published in 1841, Pugin argued succinctly and forcibly for the superiority of the Gothic. Much of the text is epigrammatic and provided a ready touchstone for like-minded, enthusiastic Gothic Revivalists. Pugin argued that the 'two great rules for design are ... 1st, that there should be no features about a building which are not necessary for convenience, construction, or propriety; 2nd, ... all ornaments should consist of enrichment of the essential construction of the building.' Ornament, Pugin emphasized, should not be 'tacked on' to buildings or 'constructed', but should be the consequence of 'the decoration of construction' – and it was, he argued, 'in pointed architecture alone that these great principles have been carried out'. According to Pugin, classical architecture was not only often deceitful – with techniques and materials of construction routinely being concealed or denied rather than honestly expressed – but also structurally primitive and limited. On the other hand, in Gothic architecture construction is not concealed but is revealed in 'truth', expressed and beautified, with the potential of the materials of construction fully realized and – through the use of pointed arches, ribs, and buttresses – given added strength through design. 'The Greeks', Pugin wrote, 'erected their columns like the uprights of Stonehenge, just so far apart that the blocks they laid on them should not break by their own weight. The Christian architects ... with stones scarcely larger than ordinary bricks, threw their lofty vaults from slender pillars across a vast intermediate space, and that at an amazing height.' So for Pugin the Gothic Revival was not just a matter of reviving Gothic forms but was about the recovery of the Gothic spirit and the emulation of authentic Gothic principles and materials of construction.

The architect George Gilbert Scott admitted in his memoirs (published after his death in 1878 as *Recollections*) that it was the 'thunder' of Pugin's writings that had first woken him from his 'slumber'. Scott, born in 1811, had to struggle hard to establish himself in the architectural profession, but by the late 1830s was starting to prosper and by the mid-1840s had gained a reputation as one of the nation's leading Gothic Revivalists. His relatively early works attempted an authentic recreation of medieval design and building techniques – for example, Exeter

College Chapel in Oxford, designed in 1854 and built of stone, with a traditionally constructed rib vault, is based on the mid-thirteenth-century Sainte-Chapelle in Paris.

During this same period Scott became embroiled in the greatest skirmish in the 'Battle of the Styles' when in 1857 he won – in somewhat dubious circumstances – the competition to build a new Foreign Office. Scott's winning design was Gothic, but he was subsequently given an ultimatum by the Prime Minister, Lord Palmerston, to produce a classical design or give up the commission. Scott, mindful of the value and prestige of the project and by nature an ambitious pragmatist, chose to shelve his principles and weather the humiliation, and ultimately executed a competent Renaissance design. While Scott was working on the construction of the Renaissance-style Foreign Office, he completed a design – in the Gothic style – for a building that is arguably his masterpiece: the Midland Grand Hotel and railway offices at St Pancras.

The designs were produced in 1865 for a splendid site on the Euston Road, just to the south of Barlow's station shed. After revisions and reductions in scale, construction started in 1868, and when completed in 1876 the building was one of the most splendid, functionally complex, and structurally modern examples of Gothic Revivalism in the world. The hotel and offices proclaimed an architecture for the age – rooted in the past, ornamented and adorned with the pedigree of history, yet functional and fit for the very modern purpose of mass travel ushered in by the new railway age. The interior of the hotel boasted – along with its elaborate Gothic-decorated public rooms – London's first elevator or 'ascending room', hot-water radiators, modern plumbing, and daringly exposed iron construction. But perhaps the true proclamation of success is that this building –

conceived as a grand hotel and after decades of office use and abandonment – is once again an elegant and successful hotel, and hugely admired for its architectural panache.

13

GUARANTY BUILDING (NOW PRUDENTIAL)
Buffalo, New York State, USA | 1896 | Louis Sullivan

The Guaranty Building in Buffalo, which Louis Sullivan designed with Dankmar Adler, is a potent and influential statement about high-rise design that was to echo down the canyon streets of New York for decades to follow. The building rises thirteen storeys high (and so is a significant development of the pair's slightly earlier ten-storey Wainwright Building in St Louis, Missouri), is of steel-frame construction, and is clad in terracotta to achieve a largely non-combustible structure.

Louis Sullivan was one of the outstanding US architects of the late nineteenth century and, along with Chicago's Burnham and Root and Charles Atwood, was a pioneer of high-rise construction. Sullivan also wrote much about high-rise architecture, so he gave the new architecture a voice, with his theories foreshadowing and defining the coming functionalist age of Modernism.

It is fascinating to relate Sullivan's writings to his key buildings, if only because they offer insights into the meaning of his architecture and the process by which it was created.

The August 1892 edition of *The Engineering Magazine* carried an article by Sullivan, entitled 'Ornament in Architecture', which was to have a profound influence on the evolution of Modernism. In the 1890s many more reflective architects were beginning to question the conventional view that, no matter how technically pioneering or innovative a building might be, historically inspired ornament was still required to give cultural pedigree. Sullivan gave these doubters a voice and a theory: 'I take it as self evident that a building, quite devoid of ornament, may convey a noble and dignified sentiment by virtue of mass and proportion … it could be greatly for our aesthetic good if we should refrain entirely from the use of ornament' and produce buildings that are 'comely in the nude'.

But despite this radical statement Sullivan made it clear, in the same essay, that he was not against all ornament; it depended on how the ornamentation related to the primary structure of the building, and the associations it evoked. A W N Pugin, in his inspirational *The True Principles of Pointed or Christian Architecture* of 1841, had thundered that 'all ornament should consist of enrichment of the essential construction of the building'. This dictate – suggesting that decoration should express or adorn structure – had clearly impressed Sullivan. He argued that, although 'excellent and beautiful buildings may be designed that shall bear no ornament whatever', ornament that was 'harmoniously conceived' and 'well considered' could be beautiful – especially when it was 'part of the surface or substance that receives it' rather than looking 'stuck on'. As for the inspiration for the ornament, Sullivan rejected direct reference to historic styles and, in the manner of Art Nouveau, suggested that architects 'turn to … the heartening and melodious voice [of] nature' and 'learn the accent of its rhythmic cadences'.This article appeared soon after Sullivan, in collaboration with Adler, had completed the Wainwright Building, which in 1931 Frank Lloyd Wright called 'the very first human expression of a tall steel office-building in architecture'.[7] The Wainwright is redolent of a Renaissance palazzo: a pronounced ground floor is topped with tiers of floors that are crowned by a mighty 'cornice', ornamented by swirling leaf designs – the 'voice' of nature – that embellish the surface of essential structure.

Four years later, Sullivan penned a piece for *Lippincott's Magazine*. Entitled 'The Tall Office Building Artistically Considered', and appearing in the March 1896 edition (pp. 403–09), the article possesses a powerful didactic and poetic presence that makes it read like an architectural manifesto for the coming age of utilitarian, machine-aesthetic Modernism: 'Where function does not change, form does not change. The granite rocks, the ever-brooding hills, remain for ages; the lightning lives, comes into shape, and dies, in a twinkling. It is the pervading law of all things organic and inorganic, of all things physical and metaphysical, of all things human and all things superhuman, of all true manifestations of the head, of the heart, of the soul, that the life is recognizable in its expression, that form ever follows function. This is the law.' And Sullivan later emphasized that it is this 'law' – that 'form ever follows function' – which governs most great ancient architecture. Essentially this is reflected in the aphorism of the first-century BCE Roman architect Vitruvius, who declared that architecture should reflect 'firmness, commodity and delight', implying that beauty is the result of solid construction and functional perfection.

That same year – 1896 – Sullivan and Adler completed the Guaranty Building in Buffalo. Like the Wainwright Building, the Guaranty has a 'cornice' ornamented with swirling plant forms, and is organized as three major horizontal 'zones', no doubt inspired by the tripartite nature of the classical column, with its base, shaft, and capital. In the Guaranty Building the ground floor contains shops and entrance vestibules, the floors above contain identical offices, and at the top is the 'cornice' zone that is both utilitarian and communal, originally including a barber's shop. Technology was all-important in its construction – the service core contained three elevators, the invention of the 1850s that made high-rise architecture feasible. The architects wanted the tall building to enjoy as much natural light as possible and for the interior to be easily adaptable, so the building is U-shaped in plan, arranged around a white-tile-clad court, and with open-plan floors, featuring beams supported on steel columns.

Sullivan realized that his early skyscrapers were inspirational, epoch-making architecture, with their own aesthetic. As he later stated in his *Kindergarten Chats*, a tall building 'must be every inch a proud and soaring thing, rising in exultation that from bottom to top it is a unit without a single dissenting line'.[8]

14

FLATIRON BUILDING
175 Fifth Avenue and East 22nd Street, New York, USA | 1902 | Daniel Burnham

---

The Flatiron Building in New York, designed by Daniel Burnham, was not the world's first high-rise building, but when completed in 1902 it was – at 87 metres (285 ft) – among the tallest buildings in the world. It was also the first to realize the artistic and urban potential of a high-rise structure built in the round – free-standing and offering a tremendous sight when seen from afar, as well as offering unparalleled prospects to all its occupants. The Flatiron was, if not the first high-rise, the world's first tower block and first free-standing skyscraper. Earlier high-rise buildings, such as Sullivan and Adler's 1891 Wainwright Building in St Louis and their 1896 Guaranty Building in Buffalo, were conceived as tall, palazzo-style structures rather than detached towers, while the Monadnock Building of 1891–93 by Burnham and Root and the Reliance Building of 1890–95 by Burnham, Root, and Atwood, both in Chicago, each form a part of an urban block and physically adjoin neighbouring buildings.

The Flatiron Building was designed to be different, and in the process it changed the nature of high-rise architecture and of New York. The opportunity was presented by the 1892 change in the New York building code that removed the obligation to use masonry as the prime structure for reasons of fire prevention. This opened the way for the use of metal-frame construction – and the George Fuller Construction Company, experienced in the construction of metal-framed industrial structures, rose to the challenge. It commissioned Burnham as designer of this office tower on the strength of his high-rise Chicago buildings (Root had died suddenly in 1891) and because his reputation was high after he had masterminded the creation of Chicago's 1893 World's Columbian Exposition, one of the most successful international fairs ever held. The Fuller Building, soon to be universally known as the Flatiron Building because its elongated triangular shape made it look like a giant iron, was to build on these successes.

The first key point was the site. Early high-rise buildings in New York had been crowded into or near the Downtown Financial District. The Flatiron was among the first high-rises to arrive in Midtown and be built not in a commercial district but in a neighbourhood with social cachet and glamour – an area where high society rubbed shoulders with showgirls, where publishing magnates and patrons of the arts met to do business and to relax. This was to start a shift in the gravity of the city and, a few years later, led to much debate and disquiet about the high-rise revolution overtaking New York and the way towers were transforming the nature of the city and the quality of life within it.

In 1902, however, the arrival of high-rise architecture north of 14th Street was more a novelty than a serious cause for alarm. The oddity of the site acquired by the Fuller Company did much to give the Flatiron its distinct character. The site, bounded by Broadway, Fifth Avenue, and 22nd Street, is wedge-shaped – an usual thing for Midtown and Uptown Manhattan, but the result of the way in which Broadway, an essentially seventeenth-century route, was accommodated within the 1811 Manhattan right-angular grid, which it crosses in a generally diagonal manner.

The site was constrained and, in a conventional sense, awkward, but Burnhan rose brilliantly to the challenge and turned a potential problem into a huge advantage. The island site meant the building could rise – in a most sculptural way – as a free-standing tower, while the shape of the site offered distinct possibilities for a theatrical composition, including an acute, prow-like corner. Also, the visual nature of the building changes, chameleon-like, when it's viewed from different angles, from broad and blockish to pencil-thin. And, of course, windows on all sides and a slender and triangular plan mean that the interior enjoys a lot of natural daylight.

From an engineering point of view, Burnham could attain the height required because he had refined the technology of wrought-iron and steel-frame construction with his Chicago buildings. As for the look, Burnham also had that mastered. His Chicago buildings had included load-bearing brick walls and fireproof terracotta panels used in a manner that was essentially simple and free of historically inspired ornament. But the success of the Columbian Exposition – for which Burnham had utilized a classical theme to give Old-World 'class' to this New-World extravaganza – had changed the national perspective. For the Flatiron, Burnham chose to hide the steel frame behind a façade of Renaissance-style classical design, organized on the established column-like principle with a visual base, then a 'shaft' of commercial space, topped in a capital-like fashion by a palazzo cornice.

Despite this deference to history, however, the building was intensely modern. The Renaissance façade was realized in fireproof moulded terracotta, with limestone used to clad the lower storeys; the twenty-two-storey building was fitted with six rapid-running Otis hydraulic elevators and had its own steam and electric plants furnishing heat and light to tenants free of charge; internal woodwork was 'fireproofed'; and floor plans were flexible or tailored to the requirements of individual occupants.

Most revealing was the public and professional response to the building. This alone made it clear that an architectural threshold had been crossed and a new world glimpsed. The

Architectural Record observed in 1902 that 'it seemed that there was nothing left to be done in New York, in the way of architectural altitude ... But the architect of the Flatiron ... has succeeded in accomplishing that difficult feat. This building is at present quite the most notorious thing in New York, and attracts more attention than all the other buildings now going up put together.' The same year, the *New York Tribune* recorded that 'since the removal last week of the scaffolding ... there is scarcely an hour when a staring wayfarer doesn't by his example collect a big crowd of other staring people ... No wonder people stare! A building 307 feet high presenting an edge almost as sharp as the bow of a ship ... it is well worth looking at.'

In 1904 Underwood and Underwood, an avant-garde photographic agency, declared of the Flatiron that 'the architectural audacity of its construction is something unparalleled in history', while in 1906, in *The Future in America: A Search After Realities*, H G Wells admitted: 'I found myself agape, admiring a sky-scraper, the prow of the Flat-iron Building ... ploughing up through the traffic of Broadway and Fifth Avenue in the afternoon light.' The architectural photographer Alfred Stieglitz, who made some of the earliest and most striking images of the Flatiron, was most succinct: 'it is the new America. The Flat Iron is to the United States what the Parthenon was to Greece.'

The Flatiron did indeed, from an architectural point of view, embody the soul of the nation and represent things to come.

15

WOOLWORTH BUILDING

233 Broadway, New York, USA | 1913 | Cass Gilbert

During the first decade of the twentieth century, the building world of New York became characterized by a collision of architectural visions. There was skyscraper mania on the part of avant-garde architects and their ambitious commercial clients, inflamed by such projects as the Flatiron Building, completed in 1902. This enthusiasm was set against a growing popular reaction that wanted the city generally to remain traditional and low-rise in character, with commercial uses kept subservient, in many city-centre locations, to established residential use. It was within this conflictive atmosphere that the Woolworth Building was created.

Although generally admired when completed in 1902, the Flatiron Building eventually provoked a reaction against high-rise office towers. Gradually people started to realize that, if it was indeed a harbinger of the future, the Flatiron promised radical change for New York – and not necessarily for the better. A New York dominated by tall towers of speculative offices would cast the city in shadow, swamp the public transport system and services, and transform neighbourhoods from residential enclaves into crowded commercial quarters, all for the sake of private profits for a few entrepreneurs. Henry James captured the mood of many New Yorkers, and anticipated evolving sentiment, when in 1907 he wrote in *The American Scene* that 'skyscrapers are the last word of economic ingenuity', ruefully lamenting a nation forever transfixed by 'the thousand glassy eyes of these giants of the mere market'.

In 1908, the newly founded Committee on Congestion of Population in New York decided enough was enough. The Singer Building, forty-seven storeys high and reaching a height of 186.5 metres (612 ft), designed by Ernest Flagg, had just been completed; the Metropolitan Life Insurance Tower, 213 metres (700 ft) high and designed by Napoleon LeBrun & Sons, was nearing completion; and now plans were unveiled for the 277-metre (910-ft) Equitable Life Assurance Building on Broadway, designed by Ernest R Graham, who had worked for Burnham on the Columbian Exposition.

The Committee proposed that New York should revise its building codes in order to limit skyscrapers, and should consider an absolute height restriction. It also recommended limiting tall buildings to certain districts, or imposing a special and prohibitive tax on skyscrapers.

These conservative views did not prevail. How could they? New York was the rapidly expanding financial heart of the nation, and the fact that the city was constrained by being located on a rocky island made the rise of the tower inevitable. If the city could not expand sideways, it had to go upwards – and the granite bedrock of Manhattan made this relatively easy.

Even if the Committee did not stop the Equitable Life Assurance Building, it did influence the construction's form – and indeed the form of the next generation of New York skyscrapers. The building was reduced in height to 164 metres (538 ft), but when completed in 1915 it was still the largest building in the world in terms of volume, and the way it blocked light sent a chill through New York. It was resolved that such a building would never again be constructed in the city, and in 1916 the first zoning ordinance was passed, which among other things controlled the bulk, and thus the design, of skyscrapers. The ordinance stated that a skyscraper's total floor area could not be greater than twelve times the size of its plot area – and if a developer wished to build higher than twelve storeys, he had to build on only part of the potential building plot, or reduce the areas of the floors as the building got higher. This led to the creation of the tiers of 'setbacks' that give many New York skyscrapers of the 1920s and 1930s such a distinct form or profile. But this legislation – and its architectural consequences – came a few years after the completion of the epoch-making Woolworth Building.

Frank Woolworth had made a vast fortune through his nationwide economy stores. In 1912 he commissioned Cass Gilbert as architect on the Woolworth Building, and by 1913 the 241-metre (790-ft) high masterpiece was completed. It has a steel frame clad with terracotta in the then traditional manner, but the organization and detailing of the building were novel at the time. The building occupies its entire site and is, like the Flatiron, free-standing and with a street-front presence. But its lower half is a broad block – U-shaped in plan to allow all offices to enjoy natural light – and from this base rises the slender tower. And rather than being Renaissance in its detailing like the Flatiron, the Woolworth Building is Gothic. Although unusual, this was a sensible choice on Gilbert's part. If historic ornament is to be used, then Gothic is a more rational choice than classical because, as it evolved in medieval churches, Gothic is an architecture of the vertical. As Gilbert observed, Gothic gave him the 'possibility of expressing the greatest degree of aspiration ... gaining in spirituality the higher it mounts'.

16

AEG TURBINE HALL
Huttenstrasse, Moabit, Berlin | 1908–09 | Peter Behrens

The office of Peter Behrens in early twentieth-century Berlin was a hothouse for the coming era of functionalist Modernism. But Behrens, an architect and industrial designer whose belief in utilitarianism made him a pioneering proto-Modernist, was something of an unlikely hero for the new age. He came from a wealthy, middle-class, and conservative Hamburg family, and in 1899 had joined the Darmstadt Artists' Colony and embraced Impressionism and Art Nouveau. By 1908, however, he had moved inexorably towards utilitarianism and a 'total' approach to architecture in which all elements, including graphics and decoration, were integrated. His office attracted – and no doubt galvanized – young men who were to play epic roles in the twentieth century. In 1908, Walter Gropius, who was to occupy a central role in the development and function of the Bauhaus, joined Behrens' office, which between 1907 and 1912 also played host to Mies van der Rohe and Le Corbusier.

When Gropius joined the office, Behrens' major project was the vast Turbine Hall for AEG, the Berlin-based German electric company – a leading generator of the power source of the age. The hall, conceived in an extremely abstracted and stripped-classical style, was an architectural pioneer. Designed in 1908 and constructed in 1909, the building anticipated many of the tenets that, in the coming years, would be used to define Modernism, which in 1908 was the term that had recently been coined to characterize the art and architecture of the burgeoning industrial age.

For the AEG building, Behrens used cutting-edge technology to achieve a vast open-span hall – 25.5 metres (84 ft) wide and 123 metres (404 ft) long – constructed using concrete panels and incorporating a wall of huge windows, with glass set in slender steel frames supported by steel stanchions that rise on 'hinge' joints. This wall of glass lights the hall in a fantastically dramatic manner, allowing machines and work areas to be illuminated with almost startling clarity.

The significant ornaments of the building are its materials and means of construction, and its form was dictated by its function. This was a building for machines that was itself designed and executed with the ruthless precision and practical perfection of a machine. With this structure, and with his functionalist convictions and his team of brilliant young assistants, Behrens became the godfather of Modernism.

All the young men in Behrens' office – including Gropius – learned a great deal from this project. New technology, in the service of an architect of discrimination, could achieve functional and aesthetic wonders, while history, when confronted with the demands of the coming world and the exigencies of modern technology, was of only very limited relevance. The AEG building's vaguely classical appearance may make it seem like some temple of power, but this is an almost coincidental consequence of the application of principles of ruthless utilitarianism. The essential point being made is that history is irrelevant as a prime source of architectural inspiration for functional buildings of the modern age, while superficial ornament is without merit.

17

THE STEINER HOUSE
Vienna, Austria | 1910 | Adolf Loos

The Steiner House in Vienna, completed in 1910 for the painter Lilly Steiner, is fascinating because it gives tangible expression to a number of the theories embraced by Adolf Loos, who was one of the most influential – and most idiosyncratic – architects of the early twentieth century.

The son of a stonemason, Loos was born in Brno, Moravia, in 1870, trained as an architect in Dresden, and went to the United States in 1893 to visit the World's Columbian Exposition in Chicago. He stayed for three years and absorbed the bubbling architectural energy of the powerful new nation. He admired the pioneering achievements of the Chicago School, with its early high-rise structures utilizing wrought iron, steel, terracotta, glass, and brick, and fell under the spell of the proto-Modernist writing of architect Louis Sullivan, particularly his brilliant 1892 essay 'Ornament in Architecture'.

In 1896 Loos returned to the Old World and settled in Vienna, where he started to design interiors and write on architecture and society in a most piercing manner. Presumably the aim was to draw attention to himself – and in this he certainly succeeded. His primary target became the established aestheticism and architecture of the once artistically radical Vienna Succession. He took issue with its indulgent, extravagant use of what he argued was superficial, irrelevant, and unnecessary ornament. In 1908 Loos published his memorable essay 'Ornament and Crime', which was influenced by Sullivan and crystallized the nature of Loos' complaint against the florid excesses of Art Nouveau and against all self-consciously artful architects and clients who craved 'meaningless' ornament.

Loos' essay can now be seen as anticipating an aspect of early Modernism defined by functionalism, minimalism, and the machine-age aesthetic. But Loos' own architecture reveals that this interpretation is a trifle simplistic. His target was not so much ornament itself as the type, meaning, and origin of the ornament used, as well as the architects who were so liberal in their use of what he defined as 'illegitimate' or 'degenerate' ornament, and who imposed their tastes and theories in a tyrannical manner. Loos' particular *bêtes noires* were architects Henry van der Velde and Joseph Maria Olbrich. Van der Velde had designed his wife's clothes so that she would 'harmonize' with Bloemenwerf House, which he had built for them at Uccle in 1895; of Olbrich, Loos wrote: 'Where will Olbrich's work be in ten years time? ... Modern ornament has no past and no future. It is joyfully welcomed by uncultivated people to whom the true greatness of our time is a closed book.' For Loos, such ornament – calculated to please a tasteless bourgeois society – was a waste of time, money, and material, and imposed tedious drudgery on those involved in its manufacture. 'Great' ornament was timeless and elemental, forged by the demands of use and construction.

Loos' mode of expression was pithy and calculated to mortally offend those whom it did not please. The opening passage of 'Ornament and Crime' was brilliant in its imagery and direct in its analogies: 'The Papuan tattoos his skin, his boat, his rudder, his oars. He is no criminal. The modern man who tattoos himself is a criminal or a degenerate. There are prisons in which 80% of the prisoners are tattooed ... It is possible to estimate a country's culture by the amount of scrawling on lavatory walls. In children this is natural, but what is natural for a Papuan or a child is degenerate for modern man ... cultural evolution is equivalent to the removal of ornament from articles in daily use.'

Loos' radical ethical and artistic declarations alienated him from most of the architectural profession in Vienna, and from potential patrons. This meant he continued to focus on writing, and through writing to reflect on architecture. In 1910 he wrote an essay entitled 'Architektur', in which he sought an inspiration for modern architecture and argued that it was absurd for the new 'urban bourgeoisie' to take refuge in an inappropriate aristocratic culture represented by European classicism. For Loos, pure, rational functionalism offered a better solution: 'Only a very small part of architecture belongs to art: the tomb and the monument. Everything else, everything that serves a purpose, should be excluded from the realms of art.'

In the same year, Loos designed the Steiner House in Vienna. The house makes it clear that, for Loos, not all ornament is a crime, and reveals in its essential composition the architect's dependence on the classical tradition. Earlier designs by Loos reflected his Anglophilia, represented by his dandified English manner of dress and love of the cosy clutter of the English Arts and Crafts interior. The Steiner House is more rational and initiated a series of buildings in which Loos evolved his conception of the *Raumplan*, or 'plan of volumes', which was a complex organization of internal spaces that resulted in split-level houses, notably the Müller House of 1930 in Prague.

Visually, the Steiner House appears the oddest bauble of a building. To the street it presents a single-storey elevation topped by a curious, curving metal-clad roof incorporating a single, centrally placed dormer that lights the artist's studio. The bold simplicity of the rendered, white-painted, and unadorned exterior, and the functional and asymmetrical disposition of the windows, anticipated the 'white box' architecture of the International Style by almost a decade, and was a language Loos himself developed with the overtly Modernistic and cubic Müller House. The garden elevation is three storeys – due to the fall in the level of the land – to give the house, with its semi-circular roof, a more monumental and geometric quality. Internally the house reflects Loos' penchant for built-in furniture. As he explained, 'The walls of a building belong to the architect. There he rules at will', which meant Loos believed that he, through the use of built-in furniture, had the right to create the interior and so ensure the full realization of his architectural vision.

18

THE RIETVELD SCHRÖDER HOUSE
Utrecht, The Netherlands | 1924 | Gerrit Rietveld

This house was conceived to be like no house before it. Indeed, it was intended to be as much an artistic statement as a house, and when completed in 1924 it was arguably not only the most innovative house in the world, but also the most visually radical.

The client, Truus Schröder-Schräder – a widowed and wealthy socialite with a passion for avant-garde design – initially wanted her architect, Gerrit Rietveld, to design a house without walls. What she got in the end is a 'house of planes', one of the ultimate expressions of architecture emulating an artistic movement. What she and her architect achieved, in a most adroit manner, was to transfer the imagery and ethos of cutting-edge two-dimensional abstract painting of relatively intimate scale to the realization of a three-dimensional, large-scale, inhabited, and ultimately functional object – a house.

The inspirational artistic movement was De Stijl, or neo-Plasticism, which had emerged in the Netherlands in 1917. It was promoted by Theo van Doesburg in the journal *De Stijl*, in painting by Piet Mondrian, and in three-dimensional design – initially and most famously in furniture – by Rietveld.

The movement explored and promoted an art of pure abstraction that would be applicable universally. It sought abstraction through reduction of all elements to essential forms and colours that, it argued, were embodied in the elemental geometry of the square, rectangles, and vertical and horizontal planes or lines, and in the primary colours, including black and white. The reduction of the complexity of the visible world to a limited series of geometric forms and colours gives De Stijl its distinct visual character.

De Stijl compositions comprising these prime forms and colours are characterized by a strong contrast between positive and negative and a sense of asymmetry, and are expressed best in the paintings executed by Mondrian from the second decade of the twentieth century until his death in 1944.

Mondrian defined the aims of the movement in various writings. As early as 1914 he wrote to the art critic and collector H P Bremmer, explaining: 'it is possible that, through horizontal and vertical lines constructed with awareness, but not with calculation, led by high intuition, and brought to harmony and rhythm, these basic forms of beauty ... can become a work of art, as strong as it is true.' And in his essay 'Neo-Plasticism in Pictorial Art', published from 1917 in *De Stijl*, Mondrian asserted that 'this new plastic idea will ignore the particulars of appearance, that is to say, natural form and colour. On the contrary, it should find its expression in the abstraction of form and colour, that is to say, in the straight line and the clearly defined primary colour.'

This is the artistic and philosophical context within which the Rietveld Schröder House was created. And the means by which these theories were made tangible – and workable for Truus Schröder-Schräder and her three children – remain fascinating.

The process of designing a house liberated from associations with traditional domestic architecture was predictably difficult, particularly since the site for this revolutionary creation was utterly conventional: it was the end plot of a standard late nineteenth-century terrace, sharing a party wall with the adjoining house, and with roads and suburban gardens nearby.

Contemporary pioneering houses by such architects as Adolf Loos and Le Corbusier, whose Villa La Roche had been designed in 1923, no doubt provided elements of inspiration, particularly for the Schröder House's free-form, open-plan first floor – but essentially the Schröder House remains an idiosyncratic one-off. Tantalizingly, it suggests a direction that twentieth-century Modernism might have gone in, although the building's eccentric character and highly personal nature make it an unlikely 'universal' model.

In plan the house is unconventional for its time, but hardly exceptional nowadays. The ground floor contains a kitchen, bedrooms or workrooms, and a central staircase leading to a first floor that is conceived as a flexible space. Only the bathroom and WC have fixed partitions, so most of the first floor can be used as one large space. However, various permutations are possible, including division of the floor into three bedrooms, by means of sliding and rotating partitions.

The exterior elevations are unconventional and exceptional. Client and architect agreed on the fundamental idea that the elevations of a De Stijl house should be pretty much a three-dimensional representation of a Mondrian De Stijl painting. The elevations are composed of a collage of planes that overlap and glide over each other, and that frame, embrace, or act as supports for the horizontal lines of the projecting balconies.

Vertical lines, an essential component of De Stijl composition, are provided by slender posts that help support the balconies and projections of the building's flat roof. The edges of these balconies and roof projections also provide additional horizontal lines. The vertical and horizontal lines are painted in strong primary colours, notably red and yellow, while the planes are painted in neutral white and grey. The client's desire to break down the division between the house's interior and exterior is reflected in the decision to continue lines, horizontal planes, and colours without regard to location, so when windows are open it is hard to tell where the façade of the house stops and the interior begins. Departing from convention, the mass of the building gives way to seemingly floating planes, to volumes that appear to penetrate one another, and to lines of vibrant colour.

In its setting the Schröder House is strange – an end-of-terrace dwelling that most determinedly has nothing to do with its homely neighbours. True, it is small, but its artistic aspirations are very big indeed.

19

THE BAUHAUS
Dessau, Germany | 1925–26 | Walter Gropius

The Bauhaus, near the banks of the River Elbe in Dessau, Germany, is a spectacular piece of construction. The main building – the nucleus of the school – was completed in 1926, but it has recently been repaired and looks not just bright, dashing, and modern, but also futuristic; not bad for a building that is nearly 100 years old. And near the main building are other structures, some designed slightly later and a few recently recreated, that reveal that this was once one of the most avant-garde creative centres of the modern world.

The Bauhaus was a school of design, but it was also an institution that embraced and promoted a specific social, artistic, and political vision of the future. It envisaged an egalitarian socialist state in which good design was seen as essential for mankind's well-being. No only were better, healthier homes and cities for workers to be created, but also all the objects needed in the home and in the city were to be rethought from first principles. All was to be related, by reference to shared design principles, in order to create total and integrated works of architecture.

When the buildings at Dessau were constructed, the Bauhaus was not a new institution, nor were its ideas entirely novel. The Bauhaus had its roots in the arts and crafts school founded in 1906 in Weimar and directed by the progressive architect Henry van de Velde. Forced to resign in 1915 because he was Belgian and thus an enemy alien, van der Velde recommended that the school be taken over by a German, Walter Gropius. But the First World War, mounting in destructive ferocity, meant that the school became dormant and Gropius did not take up the directorship in Weimar until 1919. Having served during the war (as a sergeant major on the Western Front), Gropius was well aware of the evils that result when pioneering technology is misused by man. This sombre lesson of the war, and the loss of innocence of the new machine age, had a great influence on Gropius and impressed upon him the urgency of the need for radical social change. So, under Gropius, the Bauhaus was driven by social purpose and by the determination to create a better world, with healthier and more humane buildings, for the welfare of the common man.

Although these aims had various expressions, one thing soon became absolutely clear: at the Bauhaus, the unthinking reference to history and past styles was over. Instead, beauty and meaning were to come from fitness for purpose, from the honest expression of materials and means of construction, and from the function that the object – be it a building or a chair – was to fulfil.

As Gropius explained in an early article on design: 'All inessential details are subordinated to a great, simple representational form which finally, when its definitive shape has been found, must constitute the symbolic expression of the inner meaning of the modern artefact ... automobile and railroad, steamship and sailing yacht, airship and aircraft have, through form, become symbols of speed ... In them technological form and artistic form have become a close organic unity'.[9]

If form was to follow function, and the objective and practical laws that governed the machine were to be applied to architecture and the crafts, then art had to become more scientific, and well-organized research was essential. It became necessary to consider the practical demands of the brief in great detail and to devise the most efficient way to achieve the optimum result. The elegant and minimal attainment of a design goal, utilizing the potential of new materials and means of production, became the ideal.

Gropius' *Bauhaus Manifesto* of 1919 reveals how this ideal was to be achieved and leaves no doubt as to the utopian vision of the institute. 'The ultimate aim of all creative activity is a building! ... Architects, painters, and sculptors must once again come to know and comprehend the composite character of a building ... Let us then create a new guild of craftsmen without the class distinctions that raise an arrogant barrier between craftsmen and artist! Together let us desire, conceive, and create the new building of the future, which will embrace architecture and sculpture and painting in one unity and which will one day rise towards the heavens from the hands of a million workers like the crystalline symbol of a new faith.'

Gropius turned what had started as a craft school into what was essentially the powerhouse of an art movement with a social and artistic mission that became one of the cornerstones of Modernism. The aim, quite simply, was to use art and architecture to create a better and more egalitarian world.

Gropius' early years as director of the Bauhaus were difficult, and by 1924 the institution was obliged to move to the industrial and socialist-sympathetic town of Dessau. Within a year of moving, the Bauhaus occupied a building purpose-designed by Gropius.

The structure that Gropius designed in 1925 is a powerful architectural statement – a tangible demonstration of the design principles of the Bauhaus school and a glimpse into the future of architecture. In its design, the building is a direct expression of its construction and its purpose, and it realizes the potential of new materials. Its large studio windows are filled with huge areas of glass and its reinforced concrete-frame structure is used to create open-plan and flexible interior spaces. All overt reference to history is gone, all ornament is an expression of function, and everything is doing a practical job. The interior illustrates the theory of total design. Everything is considered and coordinated – lights, railings, and furniture all speak the same bold, functional language.

In April 1928 Gropius resigned as director and was replaced by Swiss architect Hannes Meyer, whose Marxist convictions put the school increasingly at odds with growing Nazi power. In 1930 Meyer was replaced by Ludwig Mies van der Rohe, but despite his attempts to de-politicize the school, the Nazis closed the Bauhaus in 1932. It reopened in Berlin but in April 1933 was once again closed by the Nazi authorities, who declared the Bauhaus 'one of the most obvious refuges of the Jewish-Marxist conception of "art"'. In 1937, Mies fled to the United States. Gropius soon followed, leaving for England in 1934 and also arriving in the United States in 1937.

The Bauhaus had lasted only fourteen years, but its influence was, and remains, massive. It has become part of the culture of our time – artistically, socially, and politically – and has played a major role in forging the physical world in which we now live.

20

THE LOVELL HEALTH HOUSE
Dundee Drive, Los Angeles, California, USA | 1927–29 | Richard Neutra

California, particularly in and around Los Angeles, became something of a wonderland for Modernist architecture from the early 1920s onwards. This was the result of a critical combination of circumstances, but was primarily due to the arrival of two brilliant young Austrian architects, both trained in Vienna and tried and tested by their work in Europe and Chicago. These men – who, although professional rivals, would become close friends – were Rudolph Schindler and Richard Neutra.

Rudolph Schindler, born in Vienna in 1887, studied at the Vienna University of Technology and the Academy of Fine Art, from which he graduated in 1911. During his studies he came under the influence of the ideas of Adolf Loos and Otto Wagner. In 1914 he moved to Chicago, then a hothouse of contemporary design, where he worked as an architect, and from 1919 he worked for Frank Lloyd Wright, who in 1920 summoned him to California.

Richard Neutra, born in Vienna in 1892, also studied at the Vienna University of Technology. In 1921 he worked for the pioneering Modernist Erich Mendelsohn, and in 1923 – at the invitation of his university friend Schindler – he emigrated to the United States and was soon, if only briefly, in the office of Frank Lloyd Wright.

These young men rapidly evolved an architecture that responded creatively to the idyllic California climate and often dramatic landscape, that utilized the potential offered by new building technologies, that reflected the avant-garde Modernist aesthetic they had seen pioneered in progressive projects in Europe and in Wright's office, and that – in an inspired manner – captured and reflected the aspirations of their often health-obsessed California clients. Their key projects were almost exclusively private houses, and through these they expressed the spirit of the modern, the minimal, and the 'go-ahead', in the process forging a distinct California architecture that, in a wider sense, encapsulated the American domestic dream.

The start was modest enough, though in its theory radical and trend-setting. In 1922 Schindler designed a linked pair of houses, single-storey apart from covered 'sleeping platforms' on the roof and each with an L-shaped plan, located in Kings Road, West Hollywood, Los Angeles. One was for his own occupation, and the pair became known as the Schindler House. The houses abandon conventional accommodation

– rather than containing discrete living, dining, and bedrooms the houses are open plan and were intended to function as flexible cooperative spaces in which to live and work. So unprecedented was the proposed system of construction that Schindler was initially refused a construction permit. He proposed to cast a concrete slab that, before becoming the floor of the house, would serve as a work surface on which concrete panels would be cast and then 'tilted up' to form part of the external walls. This quasi-industrial system was calculated to make construction more economical in terms of time and money.

The integration of interior and exterior space, achieved by sliding glazed screens that allowed the landscape to flow into the house, combined with the house's relaxed open plan, provided a compelling model for California homes. Schindler went on to design the Philip Lovell Beach House in Newport Beach, California, which, with its expressed reinforced concrete frame, flat roof, and elevated two-storey living room, was the epitome of cutting-edge Modernism when completed in 1926.

But it was in the following year, 1927, that Neutra would design what is arguably the finest early expression of Modernistic domestic architecture created in response to the distinct climate, landscape, and prevailing social habits of California. The client was again Philip Lovell, and this build-ing, known as the Lovell Health House, was a very direct reflection of his love of healthy living, exercise, and nature.

The house sits on an elevated site over which it enjoys splendid views. In response to the rugged beauties of nature, Neutra chose to juxtapose the wonders of minimal, industrial construction to create a home, flooded by daylight, that was an arena for hygienic and healthy living, a place in which to burnish the body and liberate and illuminate the mind.

The main structure of the house is a steel frame, making this arguably the first steel-framed house in the United States. It also reflects Neutra's fascination with the architectural possibilities of industrial production and with 'technology transfer', in which artefacts from other manufacturing processes are incorporated in architecture. For example, the windows are repetitive factory-made units, while the stairwell is illuminated by headlights made for Ford's 1927 Model A automobile.

Lovell adored the house, which was no doubt largely due to Neutra's famed skill in kindling relationships with his clients and – like an analytical therapist – getting to know them and their needs. This skill was possibly learned from observations made during his early friendship in Vienna with the architect Ernst Freud, the son of Sigmund Freud and father of painter Lucien Freud.

Schindler's and Neutra's architecture had a significant legacy, notably because their work was developed and popularized in sustained and consistent manner by the Case Study Houses. Sponsored by *Arts & Architecture* magazine, these houses – many of radical modern design – were intended to provide 'dream' homes for servicemen returning from the war. The houses were largely built in California, from 1945 into the early 1960s, with a notable example being the steel and glass Stahl House, constructed in 1960 on a seemingly impossible cascading site in Los Angeles, and designed by the former GI Pierre Koenig.

21

NARKOMFIN APARTMENT BUILDING

Moscow, Russia | 1928–32 | Moisei Ginzburg,
Ignaty Milinis

The Narkomfin apartment building, constructed on Novinsky Boulevard in central Moscow from 1928 to 1932, is in many senses the prototype – both architectural and in its social vision and intent – of Modernist public housing.

It is the ultimate model for Le Corbusier's Unité d'Habitation in Marseille, France, of 1947–52, which in turn provided the prototype for much public housing in Britain from the late 1950s into the early 1970s. The British progeny include, most dramatically, the Park Hill Flats in Sheffield, started in 1957, by Jack Lynn, Ivor Smith, and Lewis Womersley, and Robin Hood Gardens in Poplar, east London, completed in 1972 to designs by Peter and Alison Smithson.

The Narkomfin apartments are currently a very sorry sight to see – mostly empty, long neglected, and in part seriously derelict. This scene of picturesque decay is in sharp contrast with the building's rational and very precise design. The apartments consist of a slab-like six-storey main block with, on one side, long horizontal strips of window lighting apartments, and on the other, wide access galleries. Particularly striking is the narrow south elevation of the main slab, with its tiers of semi-circular balconies looking rather like gun emplacements. This tough and elemental primary geometry is the most obvious external expression of the Constructivist movement that dominated Moscow avant-garde architecture during the 1920s. Also visually striking is the array of slender columns that support the structure and open up most of the ground floor as a sort of cloister, a place of light that is in pleasing contrast to the bulk of building that sits above. This inspired touch is perhaps an echo of the pilotis (columns) that Le Corbusier used in his Villa La Roche in Paris, of 1923, and was to use in a limited way in the Unité d'Habitation. The pilotis are a reminder of the international nature of 1920s Modernism. The strip windows and pilotis at the Narkomfin are almost certainly inspired by slightly earlier Le Corbusier projects, but the relationship between Le Corbusier and the Narkomfin was symbiotic, because plans of the Moscow building informed the design of the Unité.

The main Narkomfin residential block is linked by a first-floor covered bridge to a smaller, more fully glazed communal block. This latter block, containing numerous shared facilities, was the most obvious expression of the vision of communal, egalitarian, and comradely 'collective' living that underpinned the design of the Narkomfin.

Key and influential aspects of the Narkomfin include the wide access galleries that work as 'internal streets' – a conceit elaborated by Le Corbusier at the Unité through the addition of shops and cafés. There is also the building's complicated 'section', with flats of different types entered off the access galleries, some being two-storey maisonettes, and most with a dual aspect arrangement to enjoy morning and evening light. This desire to create apartments flooded with natural light throughout the day explains the narrow slab form of the Narkomfin and its north–south orientation. The flat roof of the Narkomfin contains the decaying remains of communal roof gardens, incorporating a solarium and a once-splendid penthouse.

The design and construction of the Narkomfin apartments had enjoyed the direct patronage of the People's Commissar for Finance (after whose department it is named), and they were intended to operate as a 'social condenser' that, through political education and collective living, would turn disparate individuals into focused and useful Party members. But the Commissar – like the building – fell from favour, and part of the reason for this fall can be seen from the roof of the Narkomfin. The prospect is one of desolation and urban cacophony with, in the near distance, an expression of the revisionist architecture that consigned the utopian dream of the Narkomfin to ruin. Between the grim apartment blocks that now flank the site rises one of Stalin's 'Seven Sisters', those particularly artless neo-classical towers of wedding-cake form that proclaimed the official architectural taste of the Soviet Union in the 1950s and that made pioneering Modernism of the kind embodied in the Narkomfin apartments politically unpalatable to the Soviet authorities.

22

LA MAISON DE VERRE
Paris, France | 1928–32 | Pierre Chareau with Bernard Bijvoet

La Maison de Verre, or 'house of glass', is virtually invisible. This is not because its main elevation is made of glass – indeed, the translucent glass blocks possess a tangible solidity – but because it is secreted within the courtyard of a large eighteenth-century house. The Maison de Verre, created as a home and clinic for gynaecologist Dr Dalsace, is in many ways an annex to the taller, stone-faced ranges of buildings of classical design that hide it from casual view. But despite its seemingly subservient role, the Maison is by far the most important structure in the courtyard because, in its design and materials of construction, it proclaims in startling manner the coming age of architecture.

The 1920s were the heady days of early Modernism, when avant-garde architects such as Le Corbusier were coining radical functionalist axioms such as the idea that the house is a 'machine for living' – implying that domestic architecture is essentially practical, almost disposable, certainly adaptable to changes of function, and should be driven by technology and reflect the sheer functional lines of the 'machine aesthetic'. Le Corbusier had written that a house 'est une machine-à-habiter' in his *Vers une Architecture* of 1923, and in 1929 gave tangible form to the statement with his seminal, free-standing Villa Savoye, near Paris. Almost a year earlier, Chareau gave this functionalist concept an urban identity, and in the process evolved an architecture distinctly different from the 'Purist' white boxes of Le Corbusier early villas and that, in an almost uncanny manner, anticipated the rational and exquisitely constructed 'High Tech' architecture of forty years later.

The elevation of the Maison de Verre remains stunning – almost shocking. It is fabricated of glass blocks that in the 1920s were made for, and certainly associated with, public lavatories and pavement lights. The transformation of such mean objects into the major components of a high-quality domestic elevation is a daring and brilliant move that, again, foreshadows the predilection for 'technology transfer' – when objects designed for one function are used for another, often very different function – that characterizes the High Tech architecture of the 1970s.

The glass block elevation at the Maison de Verre determines the organization of space inside the house, because the 91-centimetre (36-inch) width of the glass block panels provides the controlling dimension for the interior. And the glass-block panels not only screen inner life from outer observers but also filter light in a subtle manner that – intended or not – evokes historical associations, even though all overt historic references are eschewed. The elevation is, as many observed at the time, reminiscent of the gridded paper screens found in traditional Japanese houses, and in 1920s Paris the culture and architecture of Japan, intensely fashionable in the late nineteenth century – remained an exemplary artistic reference.

The delicate, abstract industrial imagery of the façade is continued internally. Here the primary structure is wrought of industrial I-section steels – another example of technology transfer – that are made in a sense domestic by being painted red (a colour perhaps inspired by Japanese lacquerwork), with a thin veneer of slate on their flat faces. The floor is clad with studded rubber flooring (a novel material at the time for domestic interiors), and almost every detail seems to have been rethought from first principles by Chareau, with functionalism, combination of uses, and the achievement and expression of the industrial aesthetic – or chic – being primary guiding principles: balustrades double as bookcases, the stair can be stowed away when not needed, and services are exposed so that the practical is not concealed but displayed, made almost an ornament. Again, this is one of the fundamental ideas of later High Tech architecture. Also intensely modern is the use of industrial sheet metal to fabricate fixed furniture and perforated aluminium panels to screen bathrooms from general view.

The plan of the house is complex, reflecting its cramped L-shaped site and its mixed function of home and clinic. But much is open-plan, which allows space to flow and light to penetrate so that all is, in its idiosyncratic way, a demonstration of another of Le Corbusier's axioms of 1923: that 'Architecture is the masterly ... play of masses brought together in light.' The Maison de Verre is not a play of masses or primary volumes as in Le Corbusier's early villas, but one of screens and planes. However, the role of light – filtered and direct – is of critical importance for this inspirational jewel of a building, ahead of its time, packed into a small and shadowed urban site.

Perhaps ironically – or perhaps predictably – this ruthlessly machine-like house possesses a magic, a romance, and can weave a spell. As Dominique Vellay, the Dalsaces' granddaughter, explains, 'I have loved this place passionately ... with its many screens and secrets behind open and closed doors ... Sometimes I have felt the wish to leave, but I always come back.'

23

CHRYSLER BUILDING
New York, USA | 1928–31 | William Van Alen

In many ways, the decade after 1925 was the golden age of the skyscrapers. They were no longer a new building type, but skyscrapers were still novel, and the technologies of construction and servicing were developing rapidly, making it possible

to build faster and higher – ever higher. As a child of capital-
ism, and as money-making machines riding the risks and
reaping the rewards of bold speculation, it was appropriate
that skyscrapers should come of age in the 'capital' cities of
capital – Chicago and New York – and that they should be
driven by the spirit of competition. No sooner did one
skyscraper reach a daring new height than another would be
launched to go just that much higher. This quest for height
brought structural challenges and, when achieved, also notoriety – and in notoriety and its related fame and publicity
lurked diverse business opportunities and sources of profit.

The Chrysler Building, started in 1928 to the designs of
William Van Alen, rises through seventy-seven storeys to a
height of 319 metres (1,046 ft). The client and his architect had
launched the project with the aim of creating the tallest building in the world, but became increasingly worried by the
competition offered by another New York skyscraper under
construction: the Bank of Manhattan Building. Its architects,
H Craig Severance with Shrieve & Lamb, had attempted to
wrong-foot the Chrysler team by changing their plans at the
last moment to produce a building that they hoped would be
the tallest in the world when completed in April 1930.

In response, Van Alen hatched a cunning scheme. To
change the height of the main building when construction
was under way was extremely difficult, but the dimensions of
ornamental details could be changed – such as the spire
intended to top the building. So the spire of the Chrysler was
built under wraps, its height increased and kept secret, and it
was only put in place at the last moment to beat its rival.

But the Chrysler was the tallest building in the world for
only eleven months. In 1931 it was eclipsed by the Empire

State Building, designed in 1929 by William F Lamb and started on site in March 1930, which rose through 103 floors to reach the astounding height of 381 metres (1,250 ft). After years in which New York developers jockeyed for elevational pre-eminence, the Empire State became definitive and remained the world's tallest building until 1970.

More impressive, perhaps, than the height of these buildings is the astonishing speed of their construction. The seventy-one-storey Bank of Manhattan Building was completed in only eleven months, and the Empire State in fifteen months. This was achieved by processes of construction that were put on an industrial footing and that utilized production-line methods. All these 1920s and 1930s skyscrapers are of steel-frame design, with the components factory-made to precise size and delivered to site, in relays of lorries, in the correct sequence for construction. Steel would arrive and immediately be hoisted into position and then fixed into place by the use of furnace-heated rivets that expanded when hot and shrank as they cooled, pulling the joints tightly together. This was a remarkable construction method, utilizing the cool heads, bravery, and talents of fearless men who worked in four-strong gangs and used the rising steel frames as scaffolds. Such rapid rates of construction would probably be hard to match today, in a world governed by strict codes of health and safety. Many of the 'skywalkers' in the 1920s and 1930s were Canadian Mohawks, who generally possessed a remarkable head for heights.

Riveting gradually gave way to arc welding and then bolting together of structural steel as technological advances made these techniques, which were marginally safer to execute, as effective as riveting.

The external composition of the Chrysler develops the architectural theory expressed nearly forty years earlier by Louis Sullivan – that a skyscraper, like a Renaissance palazzo, should have a pronounced ground floor, containing perhaps some public uses, then tiers of repetitive upper floors suggesting a 'democracy' of interior use, and finally an ornamental top stage. As was usual at the time, the Chrysler's ruthlessly utilitarian structural frame was concealed behind handsome masonry cladding

By the early 1930s, the top stages of New York's skyscrapers had become wonderful in their diversity and exuberance – almost buildings in themselves, defining the city's skyline and proclaiming the aspirations of the city's designers and developers. Given the nature of Manhattan – with skyscrapers set not in space but rising from the street frontages of the city's early nineteenth-century grid, creating 'canyon streets' – most top stages are best glimpsed from afar, with their often rich details visible only from the top stages of neighbouring towers.

The Chrysler's crowning feature is one of the best in New York – a marriage of Art Deco and jazzy Gothic that captures exactly the distracted spirit of the age and the tastes of the architect and the building's client, the automobile magnate Walter P Chrysler. Van Alen in his youth had proclaimed, 'no old stuff for me! No bestial copyings of arches and columns and cornices! Me, I'm new! Avanti'.[10]

The chrome-and-nickel-clad spire, along with the building's breathtaking Art Deco lobby, show what Van Alen meant. The client's interests are invoked by the vast winged discs that look a little like Mercury's hat and are based on the radiator caps fitted to the 1928 Chrysler Plymouth, by the giant hub-cap motifs that adorn an upper terrace, and by the chrome eagles at the sixty-first floor that survey the city – the heart of Mammon – with a piercing and predatory gaze.

24

VILLA SAVOYE
Poissy, France | 1929–30 | Le Corbusier

The first view of the Villa Savoye is no disappointment. Famously, this building is a Modernist manifesto; as much a diagram as a piece of architecture, it fleshes out the axiomatic theories that Le Corbusier formulated during the 1920s – notably in his *Vers une Architecture* of 1923 – and that were promoted with gusto by the magazine *L'Esprit Nouveau*, founded in 1920 by Le Corbusier and the Cubist painter Amédée Ozenfant. The villa stands white and serene in a flat, green, and lush landscape, a three-dimensional object-building that appears the epitome of the 'pure' white box, stripped of ornament (history-inspired or otherwise), that in the 1920s meant Modern. Now that Le Corbusier's rampant architectural radicalism and flamboyant embrace of the avant-garde have themselves became a part of history, it is easier to place this building – intended to shock and change the architectural world – in cultural context.

The villa is of course an embodiment of Le Corbusier's 1923 declaration that a house is no more than 'a machine for living in'. But despite its machine-like credentials and Le Corbusier's rejection of history, the villa is immersed in the classical tradition. Like Andrea Palladio's mid-sixteenth-century Rotunda, it is virtually square in plan, and it is evident that proportion and the relationship between wall and window is all-important and that Le Corbusier's famed pilotis are essentially a simplified alternative to classical columns.

The villa was conceived as a weekend home, with people flitting in and out, preferably by car. Along with the airplane, the automobile was for Le Corbusier a modern wonder that proclaimed technology's ability to liberate not just the human body but also the soul. He longed for a comparable technological revolution to engulf and transform architecture as it had transformed transport. Le Corbusier's veneration of the car is embedded in the design of the Villa Savoye. One of the most striking features when arriving at the building is the convex curve of the ground-floor entrance elevation. Why is it curved in this particular fashion? To commemorate the turning circle of Le Corbusier's then favoured car, designed by technological pioneer Gabriel B Voisin.

In *Vers une Architecture*, Le Corbusier revealed his vision of an architecture that is the 'masterly, correct and magnificent play of masses brought together in light', and stressed that 'cubes, cones, spheres, cylinders or pyramids are the great primary forms which light reveals to advantage' and are 'tangible within us ... are beautiful forms, the most beautiful.' He also asked: 'From what is emotion born? From a certain relationship between definite elements: cylinders, and even floors, even walls. From a certain harmony with the things that make up the site ... from unity of ideas that reaches from the unity of materials used to the unity of the general contour.'[11]

These somewhat abstract notions were given tangible form in the Villa Savoye, as were Le Corbusier's more didactic

'Five Points of a New Architecture', which he framed in 1927 and which aimed to exploit the artistic potential of concrete-frame construction.

Firstly, there were to be pilotis (columns) that raise the house into the air, freeing the ground floor for vehicles and people. Pilotis, which became emblematic of early Modernism, made their first appearance in Le Corbusier's work in 1922 with his second prototype of the mass-produced Citrohan House. The following year he first incorporated pilotis into built work with the Villa La Roche. Secondly, there was to be a flat roof, making space for a roof garden to replace the ground lost by the construction of the house. Thirdly, pilotis were to be used throughout the house to replace structural partitions and create a 'free plan'. Fourthly, windows were to be placed as required by the interior function to create a free façade, and finally, long, horizontal 'ribbon' windows, or *fenêtre en longueur*, were to be used to help achieve a more even distribution of light. The Villa Savoye dutifully gives expression to all these points, and demonstrates that they can be followed in the most creative way. The villa also demonstrates something else: Le Corbusier's idea about the route through a house offering a *promenade architecturale*. This takes the form of a ramp that starts opposite the front door and offers spectacular views through and out of the villa as it rises to the surprisingly large first-floor salon. This is all very much in the theatrical spirit of the Baroque. In the salon, a 10-metre (33-ft) long wall of full-height glazing opens onto a terrace from which the ramp renews its journey to the upper sun terrace with its ship-like ambience.

The villa is magnificent. Never, surely, has a diagram proved so eloquent.

25

SALGINATOBEL BRIDGE
Schiers, Switzerland | 1929–30 | Robert Maillart

The Salginatobel Bridge, completed in 1930 to the designs of Robert Maillart, is arguably the epitome of the sublime structural and aesthetic possibilities offered by steel-reinforced concrete. Its design rested on the ruins of an earlier structure. In 1905 Maillart constructed a revolutionary reinforced concrete bridge over the River Rhine at Tavanasa; this bridge possessed a slender structural form, comprising a 51-metre (167-ft) arched span, which could not have been achieved in masonry, metal, or unreinforced concrete. But in 1927, during an unprecedented avalanche, the bridge was swept away. Maillart's response was not simply to blame the loss on the unexpected violence of nature but to revaluate the design, to discover how it could have been better and stronger, and to see what lessons could be learned and applied to future projects.

The following year, Maillart got the chance to put these new design and structural ideas to the test. In 1928 he won a competition to build a bridge for a difficult and remote site high up in the Graubünden canton. The design was seminal. As Maillart's biographer David Billington explains, it shows that 'Maillart's vision had shifted away from stone images' and that the reinforced concrete structure had 'emerged from its adolescence'. Maillart had finally 'achieved a design that belongs to a new world of forms' and that 'expressed the essence of the reinforced concrete bridge.'[12]

In a sense, Maillart's design for Salginatobel is a summation of all his previous bridge designs and a declaration about how bridges should be designed in the modern world, freed from the shackles of history and liberated by the potential of new constructional materials.

As early as 1900, with his bridge at Zuoz in Switzerland and his first essay on the large-scale use of reinforced concrete, Maillart demonstrated what was to become a cardinal principle in his bridge designs: that strength should come not from mass but from form, and that if the form was appropriate and finely calculated it could also be minimal and economical in terms of materials and construction time. Maillart also developed the idea of box construction, in essence a type of honeycomb structure, which was very strong and far lighter than conventional masonry construction. All elements were, in Maillart's vision, to be fully integrated – nothing was to be superfluous or mere decoration.

By the time the Salginatobel Bridge was designed, nearly thirty years later, Maillart had refined and focused his ideas to the point where he was capable of producing a bridge that was a breathtaking synthesis of rational engineering and sophisticated aesthetics – both a work of elegant engineering, eminently fit for its purpose, and a work of abstract art.

He chose a three-hinged arch design, with flexible junctions at both bases of the arch and in its centre, largely to allow the bridge to move without undue stress. He chose to give the bridge the shape of a shallow parabolic arch because of the form's inherent strength and suitability for bridge construction and because such an arch, elegant and precise, would achieve maximum visual contrast with the fragmented and rocky landscape. These practical and artistic decisions –

and the inspired application of the structural potential of reinforced concrete – give the composition a visceral and visual power that is immediately apparent. The stark white bridge leaps across the ravine in minimal manner and achieves stability without the aid of those elements associated with masonry bridges: massive abutments, spandrel walls, and heavy parapets. Seeing the bridge for the first time can be a shock; it sends a shiver through the body and conjures up a sense of awe, almost fear, for it seems too fragile to take on the forces of nature.

The prime natural forces Maillart had to deal with were gravity, acting vertically, and wind, acting horizontally. These forces shaped the form chosen by Maillart. The parabolic arch – of 90-metre (295-ft) span – provided the most efficient means of transferring the bridge's entire gravity load – the weight of the structure and the traffic – to the cliff-face from which it springs. The detailed forms necessary to achieve the required strength in the most economic manner give the bridge its abstract beauty. For example, the threat of a horizontal movement at the ends of the arch was partly countered by making the base of the arch, and the flanking cross-walls that it adjoins, flare out. The effect of this practical decision is to give the bridge an incredibly three-dimensional sculptural feel.

The bridge is enigmatic. It can be seen simply as pure and brilliant engineering, or as a superlative abstract sculptural form, almost a great work of art by accident. It is perhaps the fact that the bridge is both of these things, simultaneously and brilliantly, that makes it such a compelling and exciting structure. What is certain is that it is a structure that was, in conception, absolutely fit for purpose, and that the response to the demands of function – giving the bridge an elegance that could only be achieved through the use of reinforced concrete – has created a work of great visual beauty.

26

THE MELNIKOV HOUSE
Moscow, Russia | 1929 | Konstantin Melnikov

The Melnikov House stands pure and perfectly geometric in a large garden among rearing nineteenth-century apartment blocks in central Moscow. It is different in every way from its gloomy neo-classical or utilitarian neighbours. It is abstract in its details, its windows an array of seemingly randomly assembled lozenges set within its three-storey cylindrical form. The house is in fact formed by a pair of intersecting cylinders, one slightly taller than the other, and incorporates a flattened elevation formed of glass. This marks the main point of entry.

The house was built by the architect Konstantin Melnikov for himself and his family in 1929, when this type of radical abstract, sculptural, geometric, and often angular architecture appeared the appropriate expression of the cultural, political, and world-changing upheaval that was the 1917 Communist Revolution.

The style that Melnikov embraced and did much to further was dubbed Constructivism. As with most radical forms of architecture of the early twentieth century, Constructivism was liberated from reference to history or overt or familiar ornament, with its form and detail seem-

ingly a response to function and to the means of construction. Constructivist theory had been first expressed in 1920 in
Naum Gabo's *Realistic Manifesto* and, in architectural terms,
in 1919–20 by Vladimir Tatlin's spiralling tower monument to
the Third International, intended for a site in St Petersburg
but never built. Connections to foreign avant-garde art movements were significant, reflecting the internationalism of
Modernism and its political dimensions. Soviet Constructivist
designers were influenced initially by the Italian Futurists,
who in particular venerated the objective and functional clarity of the machine, and later by De Stijl in the Netherlands
and Bauhaus in Germany.

Trained as an architect but tending towards a career as a
painter, Melnikov had found fame in 1925 when he designed
and supervised the construction of the Soviet pavilion at the
Exposition Internationale des Arts Décoratifs et Industriels
Modernes in Paris. This was the event that launched the international Art Deco style – a permutation of the neo-classical
movement characterized by geometric details that expressed
a sleek, streamlined machine aesthetic. But Art Deco was
little more than surface decoration, and was rarely expressed
in plan or volume. Constructivism, as pursued by Melnikov,
was more daring, all-embracing, and avant-garde. Indeed, his
pavilion was immediately recognized as one of the most
'progressive' buildings in the fair.

Flushed with success, Melnikov returned to Moscow and
designed his house. Although radical in plan, the house was
accepted by the Moscow authorities as a possible standard
prototype for state housing. This is because its cylindrical
form is structurally stable, needing no buttresses; because
brick, from which it is wrought, is strong but cheap; and

because the unusual lozenge-shaped window openings can be achieved without the addition of lintels.

But the experience of the house is far from standard. Most striking is the extraordinary plan and section, with double-height first-floor living room, spiral staircase, and rooms of diverse segmental shapes, most lit by a strange array of lozenge windows. The main bedroom – also double-height – is particularly memorable, for it still proclaims the architect's odd beliefs about sleeping arrangements. Conventional furniture was rejected as unhygienic, so here, in place of beds, were hard, tomb-like plaster slabs fixed to the floor. The architect and his wife slept on one, and on the other, screened by partitions, slept their children.

Melnikov's great, if eccentric, architectural adventure ended miserably in the early 1930s when the Stalinist regime rejected the inventive avant-garde and in a perplexing manner aped the reactionary artistic programme promoted by other totalitarian regimes – notably Nazi Germany. Constructivism, with its revolutionary novelty and sense of artistic independence, had been promoted as the authentic voice of the Soviet regime by leading Communist Party figures such as Leon Trotsky. But with the expulsion of Trotsky and the Left in 1927–28, Constructivism was regarded with suspicion and increasingly portrayed as the baffling style of a discredited clique. Elements of the Communist Party had preferred 'realist' art during the 1920s, but in 1934 'Socialist Realism' became the official artistic style of the Party. As a counter-movement to Constructivism and abstract art, Socialist Realism promoted overblown neo-classical architecture and idealized realist painting with an explicit and easy-to-discern social message.

Melnikov resolved to remain true to his principles. Rather than turning into a classicist overnight to please the Party, he gave up architecture to concentrate on painting. He became something of a recluse – a forgotten man airbrushed from Soviet history – but, until his death in 1974, he managed to remain in this house that, filled with his art, had quietly decayed around him.

The long-term future of the house – a moving monument to a galvanizing moment in world history – remains in doubt. The Russian government has stated that the house, along with all of Melnikov's possessions and paintings that still remain within it, must be preserved. But the contested ownership of the house, lack of repair, and structural threats from a massive proposed shopping development on neighbouring land ensure that the Melnikov House remains on the World Monument Fund's watch list of threatened historic buildings.

27

SEAGRAM BUILDING
Park Avenue, New York | 1958 | Ludwig Mies van der Rohe

The high-rise building, as it developed in the United States during the very early twentieth century, was a response to the increasing demand for city-centre space and was made possible by rapid technological developments.

By the end of the nineteenth century it had become possible to construct entire buildings using frames of wrought iron and then steel. This meant that buildings could rise high and fast and possess open-plan interiors that were well-lit and adaptable. Combining the technology of structural frames (with metal later supplemented by reinforced concrete) with the fast and reliable elevator, electric power, and strong plate glass meant that the modern high-rise tower had arrived, in the process introducing a new way of living and working in cities. The new high-rise building type – rooted in metal-frame construction pioneered during the Industrial Revolution in late eighteenth-century Britain and first developed in the United States in Chicago in the late 1880s – developed erratically during the first half of the twentieth century, typically with metal frames clad in masonry skins of various styles, usually ornamented, as with the Chrysler Building. Arguably the skyscraper reached artistic maturity only in 1958 with the completion of the Seagram Building on Park Avenue, New York, designed by Mies van der Rohe. It was, and in certain ways remains, virtually the last word in skyscraper design.

One of the most compelling aspects of Mies' Seagram design is that he made virtues out of the constraints surrounding high-rise construction while at the same time realizing the aesthetic potential of the building type.

Mies had been at the forefront of high-rise architecture, evolving radical ideas for sheer towers of steel and glass while working in Germany before the Second World War. In the United States he developed these ideas, notably and initially at the Lake Shore Apartments in Chicago, which were completed in 1951.

But it is at the Seagram that all the aims and ideas of the urban high-rise for commercial use come together in near-perfection. The steel and glass tower possesses tremendous elegance, with its details a direct result of the demands of the material and construction techniques used. There are no overt references to history and no superfluous ornaments. Unlike earlier skyscrapers, the Seagram does not pose as an over-stretched palazzo with a top 'cornice' and spire, a central shaft of repetitive floors, and a podium. Instead it is pure reflective wall, with the entrance distinguished in the most subtle manner by a portico-like display of steel columns. This entrance is a clue to Mies' aesthetic. Direct reference to history is, of course, eschewed, as is any form of ornament, but Mies' building is nonetheless rich in historical influence – although of a highly subtle, almost abstract sort. This influence is mostly represented through the use of traditional systems of proportions, used to unite elements of the building with the building as a whole. Or, as with the portico-like columns of the porch, details allude to ancient classical prototypes.

Also temple-like, but in response to planning laws, the tower is set within its own small square, free-standing to be seen in the round, offering wonderful views, and able to enjoy maximum natural light.

In Chicago Mies experimented with cladding the steel frame within a curtain wall of glass. Having no structural role beyond supporting its own weight means that a curtain wall can have minimal structural components, such as the frames carrying the sheets of glass, leaving maximum area for the glazing. The idea was not entirely new, but at the Seagram the system was refined to produce a shimmering skin of amazing sophistication and sleekness with slender steel mullions of the smallest possible dimension.

Inside, the functional and artistic advantage of Mies' structural system is immediately clear. The interior is open, with no structural walls beyond those enclosing the central service core. Light floods in through the floor-to-ceiling glazing of the curtain wall and the space can be used in virtually any way desired.

Although minimal in detail, the interior has dignity and beauty – and this is to do with proportion. The effect is particularly obvious in the entrance hall, where Mies achieves elegance in a simple manner: the fully glazed walls offer views out and let maximum light in. As Mies said, in imitation of a line in an 1855 poem by Robert Browning, 'less is more'.

CHAPTER 2

RHETORIC (BUILDING WITH A MESSAGE)

BUILDINGS WORK IN MANY strange and wondrous ways and have a diverse range of stories to tell. They speak through the material of their construction or the manner of their design, which might express the aspirations and wealth of their makers or reflect a building's purpose or function. For example, a great fortress such as Crac des Chevaliers in Syria largely expresses its defensive and offensive strength through its design and materials and the techniques of construction.

But buildings can, of course, speak through their ornaments as well as through their specific forms. And their messages can be direct or implied. The direct variety has been christened *Architecture Parlante* ('speaking architecture'), which explains its function and identity through its design. The phrase was coined in France in the mid-nineteenth century to describe the work of such late-eighteenth-century revolutionary French architects as Claude Nicolas Ledoux, whose architecture was often intended to carry very direct messages. He adopted or adapted classical motifs with the intention of giving his buildings a voice. For example, in Ledoux's largely visionary plan for the salt-producing town of Chaux (the Royal Saltworks at Arc-et-Senans), he designed in the 1770s a Doric entrance portico incorporating a cavernous hall that evoked the atmosphere and appearance of an actual salt mine, hoop-makers' houses in the form of barrels, and a brothel that in plan was conceived as an enormous erect phallus. This was speaking architecture with a vengeance, where the messages were loud and clear and left little to the imagination. The entrance portico was constructed but sadly the hoop-makers' houses and brothel were not.

Almost as explicit in its meaning is the Tower of the Winds in Athens, built in the first century BCE, which proclaims its purpose through its sculptural decoration; while the late-fourteenth-century Court of the Lions at the Alhambra, Spain, has the name of its creator embroidered in stone, in most decorative manner, in its very architecture; and the entire design of the late-sixteenth-century Triangular Lodge at Rushton, England, is dedicated to the number three – perhaps in homage to the Christian Holy Trinity or for some now unclear esoteric purpose.

More subtle, in a way more implicit, are the various 'nonce orders' invented by architects inspired by the orders of classical antiquity. These new orders were generally intended to carry a message about a building's meaning or function. For example, the 'Delhi order' invented by Sir Edwin Lutyens for the Viceroy's House (now Rashtrapati Bhavan), built 1912–29 in New Delhi, India, was embellished with temple bells instead of Ionic volutes to represent – in miniature – the fusion of Western and Eastern culture and religions that the design of the building as a whole was intended to symbolize. A more recent example of *Architecture Parlante* is the Bibliothèque National de France in Paris – designed by Dominique Perrault and built 1989–96 – which takes the form of a series of huge slab-blocks with each pair set at right-angles to each other like vast, open books.

28

TOWER OF THE WINDS
Roman Agora, Athens, Greece | c.100–50 BCE

The Tower of the Winds stands, an object of geometric beauty and gnarled but seemingly complete, among the battered ruins of the Roman Agora in Athens. This is perhaps one of the tower's main attractions: it seems a miraculous survivor from a fabulous and almost mythic age. It still stands while most of its contemporaries are little more than rubble or memories. True, some Grecian temples survive, many of which are older, but these are great beasts of buildings – sacred and sublime, the domains of gods and once the centres of mighty cities and cults. The Tower of the Winds, on the other hand, is relatively humble in its size and function even though exquisite as a piece of architecture. It was, in essence, a clock tower and wind vane – a most important, if hum-drum, function within any society that lives and trades by time, the tides, and the weather. Time was told by a combination of sundials and a water clock, and the building's 12-metre (40-ft) height was probably intended to allow the time of day to be seen easily by all using the agora.

The octagonal form of the building reflects the fact that for the Greeks there were eight winds, and each is personified on the tower in splendid sculptural manner. The images of the winds are placed at the top of the octagon, where they form a deep and continuous frieze that no doubt was – and remains – the visual glory of the tower. The depth of the frieze is in harmony with the basic mathematics and geometry of the tower, which is a third again high as it is wide.

As well as these images of the winds – each one facing the direction from which it, by tradition, blows – there was an added detail that gave the tower a function in the monitoring of the winds and that in a sense brought the carved images to life. This now missing detail is described by the Roman architect and theorist Vitruvius, who lived from around 80 to 17 BCE, in his *Ten Books on Architecture*. This is the only architectural treatise to survive from the ancient world, so it is thrilling to see a building surviving – in reasonable condition – that Vitruvius knew. In Book One, while reflecting on the directions of streets in city planning in relation to prevailing winds, Vitruvius observes: 'Some have held that there are only four winds: Solanus from due east; Auster from the south; Favonius from due west, Septentrio from the north. But more careful investigators tell us that there are eight. Chief among such was Andronicus of Cyrrhus who in proof built the marble octagonal tower in Athens', showing 'the several winds ... and on top ... he set a conical shaped piece of marble and on this a bronze Triton with a rod outstretched in its right hand ... contrived to go round with the wind, always stopping to face the breeze and holding its rod as a pointer directly over the representation of the wind that was blowing.'[1]

When it was first studied and documented, and its form and details published from the 1760s onwards, the Tower of the Winds became highly influential. This was partly because it provided a useful prototype for an antique tower – a form rare in Grecian architecture and most desirable for neo-classical architects designing buildings such as Christian churches, where towers are virtually obligatory. Another surviving prototype, also in Athens, is the cylindrical and column-clad Choragic Monument of Lysicrates of *c.* 335 BCE. Both these structures were illustrated in 1762 for the first time in splendid detail in the highly influential *Antiquities of Athens*, compiled by the British architects James Stuart and Nicholas Revett.

Also fascinating – and revealed in lingering detail by Stuart and Revett – is the design of the capitals used on the porches once attached to the Tower of the Winds. These have an idiosyncratic form: they are essentially Corinthian, embellished with acanthus and palm leaves but with no volutes. When their design was first published in 1762, these capitals revealed that Grecian architecture could be more individual and original than had been generally assumed.

The perfection of the Tower of the Winds, promoted through publications, inspired many architects to copy it in varied permutations – particularly the column capitals, which enjoyed a vogue in late eighteenth-century Europe. This readiness to imitate models of antique excellence reveals much about late eighteenth-century and early nineteenth-century taste in Europe and America; it was an age when originality and novelty were less important artistically than propriety and the veneration and reproduction of accepted master-pieces from the ancient world. Notable copies include the Radcliffe Observatory in Oxford, of 1772, by Henry Keene and James Wyatt, which even includes sculptural images of the winds; the garden follies in Shugborough Park in Staffordshire, of 1765, and Mount Stewart in County Down, Northern Ireland, of 1782, both by James Stuart; and the octagonal and colonnaded tower of the 1819 St Pancras New Church, London, by William and Henry Inwood, which combines the inspiration of the Tower of the Winds with the Choragic Monument. In the United States, the Merchants' Exchange in Philadelphia of 1832–34, designed by William Strickland, is crowned by a near perfect copy of the Choragic Monument.

29

CRAC DES CHEVALIERS
Syria | c.12th–13th century

When T. E. Lawrence – the future 'Lawrence of Arabia' – visited Crac des Chevaliers in 1909, he concluded that it was 'perhaps the best preserved and most wholly admirable castle in the world'. And Lawrence was in a good position to know. During the previous couple of years, while still an undergraduate at Oxford, he had made a pilgrimage, largely by bike, to the great castles of England, Wales, France, Syria, and northern Palestine for a thesis exploring the influence of Crusader castles on European military architecture to the end of the twelfth century.[2] Lawrence was correct in his appreciation of

Crac. As a later authority has written, the 'whole structure is a brilliantly designed and superbly built fighting machine'.[3]

The castle as it now exists dates essentially from around 1170, with major additions and reconstructions of the 1180s and from the first decade of the thirteenth century. It epitomizes the then current thinking about the science of military defensive and offensive design.

A strong castle like Crac could command an entire region, and thus for over a hundred years Crac played a key role in maintaining the security – and ultimately the very existence – of the Christian Crusader states in an alien and increasingly hostile terrain. When King Andrew II of Hungary visited the castle in 1218 he called it, rightly, the 'key of the Christian lands'.

The castle occupies a strong natural site made stronger by design, with defence in depth achieved by concentric rings of walls. There is a lower outer curtain wall forming the outer bailey, strengthened by closely spaced half-round towers. This defensive structure is overlooked by the far higher walls of an inner curtain wall that incorporates a series of immensely strong towers, and on its west and south sides is a steep glacis, or ramp, intended presumably to safeguard the wall from earthquakes and from being battered by rams or approached by siege towers. The glacis was a key component of the castle's system of defence. Most castles that fell to an attacking force were either starved into submission through siege or were fatally weakened by a section of wall being undermined or battered down. The fortifications at Crac are built on rock, so undermining would have been near to impossible, but battering by rams or catapults was still a threat. To help prevent such assaults it was essential to keep attackers from the bases of the walls and towers, and the glacis was a most effective way of doing this.

Another cunning defensive measure includes a tortuous route of approach from outer to inner bailey – essentially a vaulted, switch-back ramp, with a roof pierced with 'murder holes' through which missiles could be hurled at attackers, and walls furnished with fighting stations. During a direct assault this roofed passage would for attackers have become a frightful killing ground.

The inner curtain wall frames the courts of the inner bailey, onto which front the castle's most significant buildings, such as the great hall of the knights and the chapel. The largest and strongest towers are placed at the south end of the inner bailey and interconnect to form a vast donjon. This was a response to the fact that the south side of the castle site is its most vulnerable, with no natural defences.

The rounded towers on the outer wall, and on the west side of the inner wall (where their cylindrical forms rise in thrilling and sculptural fashion from the diagonal glacis), were pioneering when built in around 1205. Previously towers tended to have right-angular corners, but round towers were superior designs because they were better able to deflect stones slung by catapults and offered defenders better fields of fire. These towers contain arrow slits – as do the walls – to enfilade attackers with arrows and deny them 'dead ground' at the foot of the walls on which to gather. The top of the wall also possesses a rich array of fighting positions, notably cantilevered stone-built machicolations and stone corbels from which larger machicolation boxes would have projected and from which missiles, stones, and boiling liquid could have been poured onto attackers below.

Crac was created and held by the Knights of St John of Jerusalem, or 'Hospitallers', a military monastic order that were partners and rivals of the Knights Templar. The Hospitallers were founded in Jerusalem shortly after the First Crusade took the city in 1099, and the group's members, who possessed a high birth and warrior status, assumed an active defence of the Holy Land and the protection and sustenance of Christian pilgrims. Following papal recognition of the Order in 1113, gifts and revenue flowed in – to give to the Order was to lay up treasure in heaven – and it became rich and powerful.

In 1144 Count Raymond of Tripoli gave the Hospitallers a huge estate, including an ancient castle that was transformed into Crac des Chavaliers – described as 'the finest and most elaborately fortified castle in the Crusader Levant'.[4]

The castle withstood the buffets that beset the Crusader adventure, notably the aftermath of the Battle of Hattin in 1187, when Saladin and his Muslim forces destroyed the Crusader army and in three months took back Jerusalem. At this time Saladin contemplated Crac and decided it was too strong to attack. The power of the castle's design meant that it – and the land it commanded – remained in Crusader control. The importance of a great castle could not be better demonstrated.

But the fortunes of the Crusader states gradually declined, and when, in 1271, Crac was attacked by Baybers, the immensely powerful Mamluk sultan of Egypt, the defenders of the castle seem to have lost the will to resist. Baybers assaulted the vulnerable south side of the castle, and when the outer wall was breached the defenders surrendered. After little more than a month, it was all over. The knights left their great castle, never to return, and in 1291, with the fall of Acre after a bitter siege, the attempt to create a Christian kingdom in the Holy Land was over.

The siege of 1271 was not the last action to be seen at Crac des Chevaliers, however, and as recently as 2013 it suffered bombardment as the Syrian army drove out insurgents. Damage to the castle was significant, including the loss of tracery in the windows of the great hall – and while the region remains troubled, more damage is likely.

30

ALHAMBRA
Granada, Spain | c.14th century

In 711 CE, Muslims from North Africa – the Moors – invaded Spain, which was then divided among weak and often warring Christian states. In the south the Moors established the land of al-Andalus, which during the following centuries was to become one of the richest and most cultured places on earth – the promised land of Islam, the jewel of the Muslim world. In this land, hot and dry in part but also fertile and cool in its highlands, a distinct and rich civilization took root and flourished, despite warfare with Christian neighbours and strife within the Islamic community. Under Muslim control, but with the occasional participation of Christians and Jews, al-Andalus became a paradise of learning, with its cities like Cordoba and Granada achieving world fame as centres of pioneering medicine, science, philosophy, and the arts – particularly architecture. And notable among the architecture of al-Andalus was, and still is, the Alhambra. Like many great Islamic buildings, the Alhambra was in many ways a secret world, with an intense and complex inner life – portions are veiled, and many functions discreet and segregated. Also like many Islamic buildings, it was inspired by the desire to create a semblance of paradise on earth.

The walls protecting this hidden world are high, and the site extensive – a small fortified town rather than a palace. The most memorable way to enter this large and complex structure is by means of a disarmingly small gate that leads, via a narrow court, to the magic world of paradise gardens and celestial domes that the Nasrid dynasty, notably Yusuf I, Sultan of Granada, created during the 1330s within the central portion of the Alhambra.

The entrance court leads first to the Comares Palace, built for Yusuf I and his successors, which contains the Mexuar – a large, low room used for counsel with ministers. At the far end of this room is a glorious and richly decorated little oratory where, sitting on the floor and looking through large windows at the dramatic landscape, men would pray and contemplate the wonder of nature, of God's creation.

The Mexuar marks the start of a sequence of rooms and external spaces that reveal how the palace functioned during the fourteenth century. The route from the Mexuar leads to the Patio del Cuarto Dorado. At one end of the patio sat the sultan on his throne, giving audience and dispensing justice. The doors behind the throne open onto two routes, both of which in the days of the sultans were only to be trod by the elite of the land.

The door to the sultan's left led to his private apartment, while that to his right led into the heart of his palace. To be invited to pass through this door was to be invited to enter paradise. It leads to the Court of the Myrtles, a great reception room that is open to the sky – the heart of the palace, and a place of breathtaking beauty. There is a court, of double square proportion, in which the works of man complement the works of God. Plants, including myrtles, line the walks; to Islam, nature was sacred, a gift of God and a sign of his greatness. As God says in the Koran, 'We send down water from the sky with which We bring forth the buds of every plant ... Surely in these there are signs for true believers' (6th Sura, al-An'am). To reinforce this connection between God and nature there sits, in the centre of the court, a large, rectangular pool of water. Glassy, still, and clear, this water reflects the sky above and the buildings on each side, and at night the stars and the moon. The water is like a window into another world, an image that echoes the symbolism of the six-pointed star – the Seal of Solomon – that, with its equilateral triangles pointing up and down, proclaims that the world above regards that below, declaring that what you do here will be judged in heaven.

Beyond this court is the yet more exquisite Palace of the Lions. It was built for Muhammad V between 1362 and 1391, and is the apogee of the high artistic taste and abilities of the Nasrid dynasty. It incorporates a hall embellished with *mocárabes* – delicate, prismatic, stalactite-like forms that enliven vaults and arches – and the Court of the Fountain of the Lions. This court has four channels of water running through it and joining the central fountain, with each channel marking a cardinal point on the compass. The Garden of Eden was said to have been watered by four rivers, and this court represents in stylized miniature the Islamic vision of paradise. There is also another meaning: the circular basin, supported by twelve lions, is no doubt reference to the 'sea of bronze', supported on twelve bulls, that the Old Testament describes as being in Solomon's Temple in Jerusalem, that structure constructed under divine guidance. So this palace, a place of beauty, was also to be a place of spiritual power and an evocation of the Temple in Jerusalem that was sacred to Jews, Christians, and Muslims, suggesting that its creator was the new Solomon. The Temple connection is continued in the adjoining Sala de los Reyes – which was open only to the sultan and the elect of his court – because the 3:1 proportion of the Sala is the same as that, according to the Old Testament, of the 'House of God' in Jerusalem (II Chronicles 3: 3).

To wander the courts of the Alhambra is a pilgrimage in which beauty and meaning coalesce in mesmerizing manner.

31

RUSHTON TRIANGULAR LODGE
Rushton, Northamptonshire, England | 1593–97 | Sir Thomas Tresham

The Rushton Triangular Lodge, constructed between 1593 and 1597, is a building dedicated to the number three. It has three elevations, each 33 feet (10 m) long, and is three windows wide and three storeys high; its façades bear three Biblical texts in Latin, each 33 words in length; there are trefoil windows, and three crowning gables over each façade; and inside there is a central hexagonal room flanked by three small triangular spaces.

It is assumed that this obsession with number three is a reflection of Christian veneration for the Holy Trinity of Father, Son, and Holy Ghost, and that the lodge is a somewhat idiosyncratic religious building. Its creator, Sir Thomas Tresham, was a Roman Catholic in an age and place when to be so was – to put it mildly – inconvenient. Before 1593 he had been jailed for his beliefs, on and off, for no less than fifteen years by the Protestant authorities, which generally held that to be a Catholic was tantamount to being a traitor, or at least a person aligned with Spain, the most powerful Catholic nation and England's enemy. So it is possible that the tripartite nature of the lodge is a declaration of Tresham's evidently strongly held religious convictions, and a play upon his own surname that can be seen as embodying the number three. But none of this is certain, and other readings of this strange and enigmatic building are possible.

The Elizabethan and Jacobean age was a period of intense and often obscure occult philosophy, as is revealed by much of the literature of the time (not least some of Shakespeare's plays, such as *The Tempest*), which centres around magic, wizardry, and ancient arcane learning. It must be remembered that Queen Elizabeth placed great store in the magus John Dee and that James I wrote serious books about witchcraft (for example, *Daemonologie* of 1597) and on the divine, almost magical nature of kingship. This belief in magic – and the power of sacred numbers – was not intended to contradict Christianity but, rather, to support its mysticism and reveal coded meanings in the Bible. As Dee noted, in his preface to Sir Henry Billingsley's 1570 translation of Euclid's *Elements*, 'by number, a way is had to the searching out and understanding of every thing'.[5]

Such Christian 'white' magic echoes the Jewish Cabala, an ancient system of belief that seeks, through investigation of the God-given Hebrew alphabet, to discover the secret knowledge enshrined in the sacred texts of the Tanakh, specifically the Torah. The object of this quest is to comprehend the majesty of God and his creation, to define the nature of the Universe and the place of humans within it, and in the process to increase the faith and self-knowledge of believers.

The leading occult and humanist philosophers of fifteenth- and sixteenth-century Europe, well known by English adepts such as Dee and Robert Fludd, included the pioneering Neoplatonist Marsilio Ficino and the humanist philosopher Giovanni Pico della Mirandola, both active in Florence in the late fifteenth century, and the Cologne-born magician, astrologer, theologian, and alchemist Heinrich Cornelius Agrippa.

Ficino's occult beliefs were ultimately based on the teachings of the supposed Egyptian sage Hermes Trismegistus (the 'thrice great' or 'thrice wise'), who Ficino believed to be a contemporary of Moses and a prophet of Christianity. The sage was conceived as an incarnation of the Egyptian god Thoth and the Greek god Hermes, who were gods of magic, wisdom, and transformation (hence their importance to alchemists like Agrippa). Thoth and Hermes were viewed as guiding spirits for the soul after death, and so were seen as ideal patrons for mystic quests into hidden worlds. Agrippa set out schemes for conjuring angels and spirits through cabalistic magic activated by the manipulation of Hebrew letters that were given numerical values. He also embraced the power of the number three, as manifested through the 'three worlds' or stages of creation, which were believed to express the fact that the Universe is divided into the material world of the elements, the celestial world, and the supercelestial world of angels and divine powers, where – according to Edmund Spensers' 1590s *Hymne of Heavenly Beauty* – the 'Holy Trinity reigns'. These works define the context within which Tresham's lodge must be considered.

The Latin texts on the lodge are ambiguous, but are no doubt clues to its meaning. One states that 'Number Three bears witness', while another is from the Book of Isaiah: 'Let the earth open, and ... bring forth salvation' (Isaiah 45: 8).

So was Tresham's lodge not just an act of Christian veneration, but an invocation of the occult power of the number three? Perhaps a creation for 'searching out and understanding', an expression of Christian Cabala, and an attempt to 'conjure' – in the manner pursued by such adepts as Agrippa and Dee – angelic beings from the supercelestial world; in a sense, a main line to Christian spiritual insight and enlightenment?

32

OLD SHOIN (AND THE NEW PALACE)
Katsura Imperial Palace, Kyoto, Japan | c.1620–29
The New Palace | c.1654

The Katsura Imperial Palace was started *c.*1620 by Prince Hachijo Toshihito, a descendent of a powerful imperial family and a man with a profound romantic streak. The landscape and buildings were inspired by the eleventh-century *Tales of the Genji*, written by Murasaki Shikibu – arguably the world's first 'psychological' novel, full of emotional and sexual complexity, and partly set at Katsura. But more significantly, Prince Hachijo sought to create a landscape and buildings at Katsura that reflected Zen Buddhist theories of *wabi-sabi*, a Zen Buddhist aesthetic based on the acceptance of the transience, imperfection, and incompleteness of human existence. These three states can be reflected – in architecture and landscape design – by asymmetry, asperity, simplicity, economy, modesty, and integrity.

The traditional Japanese house strikes a fine and distinctive balance between design and construction. Design – the ways in which spaces are conceived, created, and used – is of crucial importance, yet the material realization of the design can be slight and subtle, making the Japanese house a triumph of the minimal, the ephemeral, and the flexible. It is also a meticulous machine in which strict ritual is observed in ways so subtle as to be (initially) imperceptible to those not initiated into the culture.

The Japanese house is, by its nature, a lightweight structure utilizing readily available traditional building materials. Almost invariably timber is used for the main structure, with walls made of timber lattice screens of translucent paper (*shoji*). These screens can be fixed, or they can slide (*fusuma*) and be used to manipulate space. Sometimes the screens incorporate paper embellished with ink paintings (*haritsuke kabe*). These sliding screens, used for elevations and internal partitions, help to break down the boundary between interior and exterior by allowing external walls to expose and frame views of the surrounding landscape.

In this architecture the house is conceived as part of its setting and its structural strength comes not from mass or materials but from ingenuity of construction. In such architecture, meaning lies not in the material aspect of a building – in its materials – but in the spiritual ideals it represents.

Inspired by such concepts as *wabi-sabi*, the traditional Japanese house is, as in many Asian cultures, more than just a home. It is also a sacred building, often partly and literally a shrine, reflecting religious beliefs and philosophical attitudes. These include not just Buddhist doctrine but also Shinto, the indigenous religion of Japan. Shinto venerates the sacred and omnipresentforces of nature, which are thought to permeate matter and space and are made manifest in natural creations such as trees, rivers, and stones, rendering each sacred. All these ideas find expression in architecture, notably in materials used, design, site, and orientation.

To talk of the traditional Japanese house is, of course, to grossly oversimplify, since there are powerful regional variations as well as differences due to scale and expense of construction and changes through time. In addition, as in most societies, houses in Japan are a close reflection of status. There were the vernacular *Minka* ('houses of the people'), which in the context of Japan's traditional 'four divisions of the people' were dwellings for artisans, farmers, and merchants. The samurai warrior caste could have very different houses, large and fortified, such as the *Shinden-zukuri* palatial architecture of Heian-kyo (now Kyoto) in the eighth to twelfth centuries, comprising separate pavilions for different uses (eventually including tea houses), with square courts as the primary outdoor space.

The traditional house includes specific rooms, types and arrangements of furnishings, and structural devices with symbolic meanings. Broadly speaking, the design and furnishing of these rooms can reflect general principles relating to patterns of life – notably the three concepts of *shin* (formal), *gyo* (semi-formal), and *so* (informal). These principles evolved in calligraphy and flower arrangement and were adopted in architecture, becoming particularly important in the design and rituals of the tea room and tea ceremony (*chanoyu*). They could also be expressed very directly through construction, with, for example *gyo* or *so* interiors containing tree-log columns with bark un-removed and walls plastered with mud.

The most important room of the house is the *zashiki*, which in larger houses is the principal reception room and living room (perhaps incorporating a shrine), and in smaller houses is simply the main room, multi-use and versatile, used for sleeping as well as dining and sitting. Inevitably the floor of the *zashiki* is covered with *tatami*, a type of mat tradition-ally made using rice straw and always of a 2:1 proportion, so two side-by-side make a square covering. This reveals the proportions of the ideal *zashiki*. These mats are, ideally, immaculately clean because they are not so much for stand-ing as for sitting on. For this reason all shoes are immediately discarded upon entering the house, usually in the stone-paved and so semi-outdoor entrance area.

So important is the *zashiki* to traditional Japanese living that myth and legend surround it. The most notable has to do with the *zashiki-warashi*, a playful, sometimes mischievous, but generally benign spirit or sprite that might occupy a house. In Japan such a guest is viewed as a blessing, but must be noticed and cared for properly. It must even be tolerated cheerfully when its child-like nature means the occasional tantrum, during which – poltergeist-like – it will throw plates or cushions about, conjure up eerie music, or leave its diminutive footprints in ash, usually over the nice clean *tatami* or *haritsuke kabe*.

Other key rooms or functions of the Japanese house – often with distinct architectural expressions – are the *shoin*, the writing room or building, and the *chanoyu* room or tea room. This room and the tea ritual became the centre of the *wabi-sabi* aesthetic – the focus on the beauty of art, pottery, calligraphy, flower arrangement, cooking, and tea that was 'the crystallization of Japanese culture'.[6]

The design, construction, and arrangement of the buildings forming the seventeenth-century Katsura Imperial Palace, and the exquisite garden within which they are placed, help make tangible many of the complex and abstract principles of the Japanese house.

The Old Shoin at Katsura dates from the 1620s and was constructed to entertain a large number of people at informal gatherings, its *fusuma* screens making the interior highly adaptable. The New Palace was created in the mid-seventeenth century, by Prince Hachijo's son, to accommodate the ex-emperor Go-Mizunoo. Like the Old Shoin it has a hip-and-gable roof (*irimoya*), as do most of the other major buildings of the palace, with the primary structure formed partly by square-section timber columns. Earlier structures have columns with bevelled corners, but by the seventeenth century this detail had been simplified.

The New Palace has raised floors and *tatami* mats arranged to define the orientation of interior space, focusing on porches – created by sliding screens – and platforms that

1 *The Step Pyramid at Saqqara, perhaps 2,650 years old, is a pioneering stone-built construction. It comprises six levels – with the seventh being the surface of the earth – and seems to have served as a tomb and memorial for the pharaoh Djoser. (Dan Cruickshank)*

2 *The exterior colonnade of the Temple of Apollo Epicurius, completed in around 400 BCE, is a fine and early example of the Doric order. It has short sides formed of six columns, though the far column here has collapsed. (Corbis)*

3 *The Pantheon's coffered dome is made of concrete that would have been poured onto wooden moulds supported by timber scaffolding. The drum from which the dome rises – and which carries its weight and resists its lateral thrust – is made of brick, clad internally with marble and stone. (Pablo Cersosimo/Robert Harding World Imagery/Corbis)*

5 *The Villa Almerico Capra is one of the most captivating and inspirational buildings of Renaissance Italy. It is a monument to symmetry, to the restrained, rational expression of the classical language of architecture, to powerful cubic and spherical forms, and to harmoniously related proportions. (Stefano Politi Markovina/Alamy)*

4 *The entrance front of the Palazzo Rucellai, Florence. Constructed in the mid-fifteenth century and almost certainly designed by Leon Battista Alberti, the elevation is one of the most sublime and satisfying examples of early Renaissance architecture. (George Tatge / Alinari Archives / Getty Images)*

8 *The interior of Dulwich Picture Gallery, showing the axial arrangement of rooms linked by simple arched openings. The space is top-lit by means of large roof lanterns. (Dan Cruickshank)*

9 *The Palm House at Kew Gardens combines the utilitarian demands of its function with a sense of form that, united to the use of pioneering materials, makes it one of the most beautiful and functional buildings of its age. (David Iliff)*

12 *The main staircase in the Midland Grand Hotel in St Pancras Station is a fantastic concoction of Gothic Revival forms and details used to enliven a symmetrical 'imperial' staircase of classical conception. The structure supporting the stone treads and landings appears to be made of cast iron, but much of the ornament is timber or plaster. (Alex Segre/Alamy)*

13 The Guaranty Building was one of the tallest commercial buildings in the world in 1896. Its three 'zones' echo a Renaissance palazzo: a podium containing entrance foyers and public or communal spaces, which supports stacks of uniform commercial space and is topped by a deep – and inhabited – 'cornice'. (Corbis)

14 *The ship-like 'prow' of the Flatiron Building is seen here from the north. (Anna Bryukhanova/istockphoto)*

15 *The upper portion of the Woolworth Building soars above Manhattan, with its fine Gothic detail making it one of the most delicate and beautiful of the city's skyscrapers. As its architect, Cass Gilbert, explained, the Gothic detail meant the building gained 'spirituality the higher it mounts'. In the background, by contrast, is the sleek form of the Empire State Building, completed in 1931. (Angus Oborn/Getty Images)*

23 *Birds-eye view of the Chrysler Building soaring up into the giddy heights above Manhattan. The Art Deco and jazzy Gothic spire is the building's most characterful external detail. It is clad – automobile-like – in chrome and nickel. The main part of the building below is clad with pale brick. (Cameron Davidson/Corbis)*

18 *The Rietveld Schröder House in Utrecht – the artistic and aesthetic theories of Piet Mondrian and the De Stijl movement made tangible and habitable. (Arcaid Images / Alamy)*

24 *The Villa Savoye, essentially cubic in volume, evokes the ambiance of Palladio's villas while also expressing – in a most tangible and dramatic form – Le Corbusier's theories about the nature of the modern home. This view shows the column-like piloti and the strip windows – or fenêtre en longueur – that he favoured. (Bildarchiv Monheim GmbH / Alamy)*

30 *The Court of the Myrtles – a place of open-air reception at the heart of Alhambra – is an evocation of paradise on earth. (Roberto Adrian/istockphoto)*

29 *Crac des Chevaliers, built from the 1170s, was called by Lawrence of Arabia the 'most wholly admirable castle in the world'. Defence in depth was provided by concentric rings of walls, with the higher inner wall and its strong round towers rising off a steeply sloping glacis. (Dan Cruickshank)*

33 *The entrance façade of Easton Neston, with its giant order of composite pilasters and central columns, is intended to give the building an antique and noble air of gravity. (Dan Cruickshank)*

35 *The breathtakingly bold entrance elevation of the Altes Museum, Berlin, in which a screen of tall Greek Ionic columns proclaims the antique pedigree of the building's design and its suitability as a repository for artefacts of ancient culture. (ewg3D/istockphoto)*

38 *Cylindrical reading rooms surmount the cubical lower podium of the Stockholm Public Library and the main door, with tapering sides in the ancient Greek manner, and a ramp approach. These details, and the building's powerful and elemental geometric forms, make it clear that its design is rooted in the classical tradition. (Folio Images/Alamy)*

39 *Fallingwater was sited to perfection by Frank Lloyd Wright and is, in its design, a fascinating fusion of architectural inspiration. In the Arts and Crafts tradition, the building utilizes local*

materials but combines them with the Modernist vocabulary of reinforced concrete and much glass.
(Dan Cruickshank)

42 Wrought of concrete, the National Parliament Building confronts gardens on one side; on the other, it rises from a lake with, across the water, brick-built cylindrical and cubical structures that are pierced by circular and semi-circular windows. All proclaim the enduring power of primary geometry. (Aopu)

44 The Mortuary Temple of Hatshepsut at sunrise, looking across the forecourt and up the ramps to the distant 'Holy of Holies'. (Dan Cruickshank)

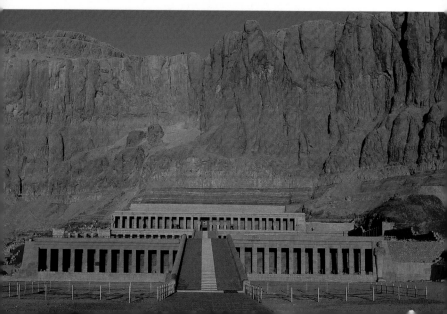

relate to the landscape in function and orientation; for example, the Old Shoin contains a 'moon-viewing' platform that is an extension of the interior space. Rooms also contain alcoves – *tokonoma* – in which objects of aesthetic or spiritual virtue would be displayed, and built-in desks – *tsuke shoin* – that give character and meaning by suggesting contemplation and refined, informed reflection.

In the garden is a series of tea houses, some with bark-covered posts that suggest informality, others with shallow windows placed to offer beautiful vistas when seated. These details, as with many others, were designed to evoke the power of the natural world and grant inner peace and tranquillity.

33

EASTON NESTON
Northamptonshire, England | c.1685–1702 | Nicholas Hawksmoor

Easton Neston in Northamptonshire is a jewel-box of a building. It was completed in 1702 and although its exact building history remains debated, it is generally accepted that Nicholas Hawksmoor was the main architect and that this was his earliest major independent commission and arguably the first building in which his idiosyncratic architectural manner came to maturity. It is also one of the finest eighteenth-century Baroque country houses in Britain.

The present house was begun by Sir William Fermor in the mid-1680s. He seems to have taken advice from Sir

Christopher Wren, who was the cousin of Fermor's second wife, and there survive an enigmatic set of drawings with enough similarities to the house as built to persuade the distinguished historian Howard Colvin to argue that the design of Easton Neston 'probably owed as much to Wren as Hawksmoor'.[7] Yet there is no evidence that Wren was formally appointed as architect.

In 1692 Sir William married his third wife, Lady Sophia Osborne. This was an extraordinary coup. Sophia was the daughter of one of the most powerful men in the land – the Marquis of Carmarthen, made Duke of Leeds in 1694. Sir William's dowry was reported to be £10,000, and within six weeks of the marriage, he was ennobled and became Lord Leominster.

Leominster now wanted a house that would proclaim his new status and it is probable that designs changed dramatically, perhaps influenced by the impressive and seemingly aristo-cratic Baroque style that had been pioneered at Chatsworth during the 1680s and 1690s for the Duke of Devonshire by the architect William Talman. Evidence for this new initiative comes in the form of a sensational model – one of the earliest architectural models to survive in Britain, and now in the Victoria and Albert Museum – which is embellished with Lord Leominster's coronet and so must date from 1692 or soon after.

The model shows clearly how the extraordinary and complex design of the existing building was achieved. The hall – double height in volume – is set at one side of the house to create an asymmetrical plan reminiscent of English medi-eval layouts, and the front door led not into this hall but into an entrance passage that ran alongside the hall towards the staircase and rear door in the manner of a medieval screens passage. Why the designer should have evoked ancient planning practice remains obscure; perhaps the intention was to flatter Leominster by giving his new home the pedigree of history and so imply ancient lineage.

The other key features shown in the model are the main staircase, rising majestically in a large volume set diagonally opposite the hall, and an ingenious system of mezzanines.

But the house as built contains one great difference from the model. Its elevations are dressed with a 'giant order' of two-storey-high composite pilasters, with the centre bay on the entrance front emphasized with a pair of full columns set in front of the pilasters. The giant order, now such a dominant feature of the house and surely part of a strategy to make it more visually impressive, was evidently a late arrival. Earlier and inspirational buildings that expressed the visual punch of the giant order – and that must have influenced the design at Easton Neston – include Michelangelo's buildings on the Campidoglio, Rome, from 1536, and the influential Château de Vaux-le-Vicomte in France, of 1657–61.

The start of Hawksmoor's involvement with the project is unclear. Most likely is that Leominster consulted his friend Wren over his choice of architect, and Wren recommended his erstwhile star pupil Hawksmoor to execute the project. A surviving letter from Hawksmoor, held in the Victoria and Albert Museum, suggests that he had been involved in the project from as early as July 1686, and Colen Campbell, in the first volume of *Vitruvius Britannicus*, published in 1715, includes an elevation and plan of Easton Neston and describes it as 'the ingenious Invention of Mr. Hawksmore', noting that it was 'finish'd Anno 1713'.

As completed, the interiors of Easton Neston offer a dramatic architectural experience in space and light. The experience would have been even more striking originally when the hall was a vast double-height cubic volume, but sadly its ceiling was lowered in the late nineteenth century to create first-floor bedroom accommodation. The entrance passage continues to the rear door, but before this is reached the main staircase is suddenly, and in most dramatic manner, encountered within its lofty compartment. This is a bold and wondrous creation, with its vast stone treads and landings suspended in space and flooded with light from a huge sash window. The staircase offers a splendid architectural promenade to the first floor, where it terminates with an Elizabethan-style long gallery that runs the full width of the house.

But when Easton Neston was completed, the staircase was more than simply the key component of a splendid architectural route through the house. It was also a gallery for the outstanding and celebrated Arundel collection of antique sculpture. This was acquired by Lord Leominster in the late seventeenth century, and he resolved to integrate the sculptures within the fabric of his new house to create an heroic and aristocratic atmosphere of scholarship and connoisseurship.

Along with Talman's west front at Chatsworth (1696) and Castle Howard (1699–1712) by Sir John Vanbrugh and Hawksmoor, Easton Neston is one of the buildings from the very end of the seventeenth century that was to launch the emotive and inventive freedoms of the English Baroque, in which large-scale orders, sculptural form, abstracted details, and antique references play a dominant role.

34

BHUTANESE FARMHOUSE
Shengana, Punakha, Bhutan | c.1750

The traditional Bhutanese house is a wonderful thing – the inspired marriage of rational construction and sparklingly bright sacred ornament.

These houses generally vary in height from two to four storeys. The lower levels are made of earth pounded into a timber mould, with the walls tied together by internal timber beams that also carry the floor structure. Using earth for the basic structure is, of course, a brilliant idea – it's freely available, does not demand specialist skills to use, and provides superb (and fireproof) insulation.

The storeys above the earth-built lower structure are timber-framed. These storeys, in which the box-frame is exposed and its panels are filled with woven bamboo covered with pine-needle-bound mud, contain the main windows of the house. These are generally fine examples of ornamental carpentry. Also finely detailed is the timber cornice that tops the house, typically incorporating tiers of square blocks that are the external and decorative expression of the joists forming the top ceiling of the house.

But the top ceiling does not mark the top of the house, because soaring above this ceiling is the house's most decorative and structurally spectacular and sculptural element. The roof structure is almost a building in its own right. It can take several forms, but essentially is built with triangular trusses strengthened with centrally placed king posts and struts and with horizontal tie beams projecting to form deep eaves all around the house.

Everything in these houses that appears ornamental is in fact a consequence of the materials and means of construction or a reflection of function – and this goes for the paint-work, too. The pounded earth walls are generally painted white with lime wash to protect them from the rain, while timber ornament is also given a protective coating of paint, with its detail coloured and picked out to reveal or underline its meaning. Colours have symbolic values, so decoration can appear somewhat abstract for the uninformed; a trifle more obvious are the phallic images painted on the external walls. These are mostly to give protection from evil, and to confer benefits on the occupants.

To explore a traditional house in detail, I went to the remote village of Shengana, in Punakha in western Bhutan, to visit a farming family in their home. The journey ended in a verdant valley with fairy-tale houses perched on each side. It seemed an enchanted land. The house I'd come to visit was isolated, tall, and seemingly ruinous, with paint flaked away to reveal naked earth walls and timber. The family told me it was between 200 and 300 years old.

The house rose four storeys, and must once have been the centre of prosperous family life. On the rear of the house was the usual two-storey timber-framed and mud-panelled veranda, starting at second-floor level. In front of the house a pair of oxen was tethered, marking the ground floor as a manger. The first floor – the location of the front door, which could only be reached by ladder (a good defensive tactic) – was given over to storage. In one corner was a steep staircase, leading to a second-floor landing; from there, a door with a high threshold and a low lintel – a defensive design, since the Bhutanese believe a spirit can't bow to enter a room – led to the main living room, a place of dining and communal sleeping where a meal was given in my honour, men and women segregated and sitting on the floor.

The third floor, reached via a steep staircase on the veranda, contained rooms where the most valuable possessions were stored, such as saddles, and where hunks of sun-dried beef – a Bhutanese delicacy – hung from strings. There was also a well-appointed Buddhist shrine. Above this floor, within the open-roof structure, were piles of corn and joints of meat curing in the breeze, and also some small clay conical objects that looked like miniature stupas. Each was made from clay mixed with the ashes of a cremated body, and so they were relics of ancestors, placed here so that the living could look after the dead – and the dead, perhaps, intercede for the living. The Bhutanese see their houses as sacred places, as miniature renderings of the cosmos – the ground floor as an underworld occupied by beasts, the middle as the world of man, and the top as heaven, the realm of enlightened beings and the venerated dead.

This is architecture without architects, without drawings or textbooks. It is the result of knowledge passed verbally from one generation of builders to the next. There is little concern for invention or novelty – a building-form has evolved over centuries that works, and all that's necessary is to keep tradition alive. And at the moment in Bhutan, tradition is still very much a living thing.

35

ALTES MUSEUM

Lustgarten, Berlin, Germany | 1825–30 | Karl Friedrich Schinkel

The Altes Museum is one of the first and greatest of the world's major public museums to be conceived as a solemn and imposing temple of art. Early structures of comparable function had either contained a mixture of uses or were far more humble in scale and in their architectural language. For example, the Uffizi, started in Florence in 1560 for Cosimo de'Medici and designed by Giorgio Vasari, contained – as the name implies – offices as well as courts, archives, and works of art. These were displayed in grandeur in the first-floor galleries of the *piano nobile* and in the large, octagonal, and double-height Tribune added in the 1580s. The Uffizi's galleries, particularly the Tribune, remained an inspiration for museum designers for centuries to come; indeed, they were one of the major influences during the seventeenth and eighteenth centuries for designers of private country houses containing galleries for the display of art. A good example is Holkham Hall, Norfolk, designed in the early 1730s by William Kent and Lord Burlington, which incorporates a gallery for the display of antique sculpture and a pair of octagonal, Tribune-like rooms for the display of paintings.

But these were private galleries. The world's first picture gallery open to the public – and independent of any other structure – was designed in 1811 by Sir John Soane in Dulwich, London. Although pioneering and architecturally sophisticated, however, this gallery was relatively small and certainly not designed to be a public temple of art.

The first major architectural expression of the conviction that great cities or nations should collect and display art – in a splendid and public manner, for the inspiration and improvement of their citizens – is the Glyptothek in Munich, Germany. This building did not usher in the notion of the state-owned public museum (the British Museum, for example, had been founded in 1753), but it did introduce the idea that such museums should be housed in purpose-designed structures that in their noble classical design, wrought of masonry and embellished with large porticoes, were to appear as temples, calculated to evoke ancient learning and virtues. The British Museum was housed for the first seventy years of its life in pragmatic manner in a late seventeenth-century urban mansion.

The Glyptothek, designed in 1816 by Leo von Klenze in a Greco-Roman style, was founded by the Bavarian King Ludwig I to house the state collection of Greek and Roman sculpture. Before this museum opened, the British Museum launched its own temple-like reconstruction. The architect was Robert Smirke, the style massively and decidedly Greek, and work started in 1823 on the King's Library. Work on the museum proper – of which the King's Library became the east range – started in 1825 and was to last until the early 1850s. By this time it had been decided that the national collection of paintings should be displayed separately and another temple-like museum was built, from 1832, on what became the north side of Trafalgar Square. Designed by William Wilkins, the National Gallery opened in 1838.

The Altes Museum in Berlin – arguably the masterpiece of the great Prussian architect Karl Friedrich Schinkel – was conceived in the early phase of this Europe-wide passion for the museum. It was built to house the Prussian art collection and was located at the end of the Lustgarten, or Pleasure Garden, opposite the Stadtschloss or Royal Palace. Schinkel started work on the design in 1822, but construction did not start until 1825. The museum opened in August 1830.

Schinkel's design is remarkably powerful. The elevation facing the Lustgarten displays an almost breathtaking simplicity of conception. It's little more than a screen of giant Greek Ionic columns rising from a terrace set on a base that is reached by a broad and noble central staircase. It's a magnificent and bold urban gesture – inspired by the temples and stoa, or colonnaded walkways, of ancient Greece – that works wonderfully well on its island site. The colonnade proclaims the building's antique pedigree and suitability as a repository of culture. It also provides a splendid backdrop to the Lustgarten, and would have formed a stunning antique prospect when viewed from the palace. Schinkel's genius was to know the right ancient model to select and how to use it; the comparable colonnade at the British Museum, by contrast, is cramped, contorted, and remote. At the Altes Museum Schinkel has brought the ancient past to life, with a colonnade that was intended to serve as a raised promenade and transform the Lustgarten into a Grecian agora at the heart of the city.

The interior, now reconstructed after war damage, holds a surprise. The external architecture is Grecian, but in the centre of the museum, flanked by long galleries, is a cylindrical and domed space inspired by the Pantheon in Rome. For this temple to the arts Schinkel has married, in almost magical manner, the essence of the two great forms of temples of the ancient world.

36

THORVALDSEN MUSEUM
Copenhagen, Denmark | 1838–48 | Michael Gottlieb Bindesbøll

This museum was built as a monument to the great Danish sculptor Bertel Thorvaldsen, who in 1838 returned to Copenhagen a national hero after achieving an international reputation during forty years' work in Rome. Thorvaldsen was seen as the successor to the great Antonia Canova and the embodiment of national pride. The new museum, essentially a hall of fame and repository for a large display of Thorvaldsen's work, including plaster casts of originals, was paid for by public and royal subscriptions. And when Thorvaldsen died in 1844 the museum became his heroic tomb, with his body interred below the central courtyard. Thorvaldsen donated his art and collection of Egyptian, Greek, and Roman antiquities to Copenhagen, so this building is not just a shrine to one man's work but also the city's first public museum.

The museum is an astonishing creation in which antique Greek imagery and inspiration are fused in dramatic manner with then modern ideas about axially planned gallery design and high-level lighting, inspired to a degree by the works in England of Sir John Soane. It is also brilliant for the manner in which sculpture and architecture are integrated and for its use of often fierce colours, thought to be an evocation of ancient Grecian practice but also powerfully contemporary. In the museum, colour possesses an emotive quality and is used to imbue Thorvaldsen's pristine, neo-classical white sculpture with warm Mediterranean hues and to set it within an intensely romantic context. The neo-antique setting, notably the bold Greek revival details, the strong earth colours associated in the 1830s with ancient Greek architecture, and the varied ceiling decorations – some recalling galleries in the Vatican, others strongly coloured barrel vaults – give the often excessively chaste sculpture an added depth of interest and brilliance.

The main entrance elevation to the museum is essentially a demonstration of the theories current in the 1830s about the way Greek architecture might have looked 2,500 years ago. It has five tall doors, with architraves that taper from top to bottom. These doors incorporate windows in their upper portions, and this addition gets around a practical problem that was most awkward for Greek Revivalists in the early nineteenth century: authentic Greek temples (which provided Revivalists with their models) rarely had windows, and certainly did not make dominant features of them.

But more striking than these five closely set and generously glazed doors are the colours used to adorn the elevation. It is painted strong yellow ochre, but the door architraves and cornices and the elevation's main entablature are picked out and coloured in a most vibrant manner, with off-white, dark blue, and dark red dominating. The top cyma moulding of the cornice is even painted with stylized floral decoration. This elevation – topped by a bronze quadriga depicting Victory – sets the archaeologically inspired, polychromatic neo-Greek theme for the building.

For those who love the neo-Greek, the building is bliss. But even those more sceptical about the artistic virtues of the movement will no doubt find the interior impressive for its determined single-mindedness, its creative invention around limited themes (every ceiling – barrel-vaulted or otherwise – is different), the sheer gusto of its colours – mauve, dark pink, Pompeian red, deep yellow ochre, dark blue and green – and the bold way in which the sculpture, often set in isolated splendour, is placed against great blocks of these colours, often in rooms that are cubical, or a permutation of a cube, in volume. And then there are the vistas along the long side galleries that frame the central courtyard. In the manner of the Baroque, gallery rooms are arranged in enfilade, with doors aligned, to offer extraordinary and kaleidoscopic perspective views.

The courtyard is a closed world, its long sides furnished with rows of tall tapering doors, their upper portions once again glazed. But there are also square attic windows set above the doors – a functionally essential but un-Greek detail – to help light the first-floor galleries. The focus of the court is the short side elevation facing the entrance range, with its single tapering door, rising two storeys and containing an upper window. Beneath the first-floor windows of the court are images inspired by attic vases, an association emphasized by the colours of the walls. But rather surprisingly, on the short wall are large and somewhat naturalistic renderings of swaying palm trees. All most exotic – particularly when viewed in Nordic snow, the intense colours burning with an ardour for antique, Mediterranean culture.

37

WHARENUI (MAORI MEETING HOUSE)
**Whangara, Gisborne, North Island, New Zealand |
1890s and 1939**

The town of Gisborne, on New Zealand's North Island, stands very near the site where Captain James Cook and his party first set foot on this beautiful and bounteous land on 8 October 1769.

The consequences of Cook's arrival in New Zealand and his possession of the land for the British Crown were far-reaching – particularly for the people who were the well-established residents of the land. But the Maori, as these people came to be called, were themselves relatively recent settlers.

There is still much debate about the origin of the Maori and exactly when and why they arrived at New Zealand. What is certain is that they are a Polynesian people and that they arrived by canoe, probably from the northeast. The consensus at the moment is that the origin of Polynesians lies in Taiwan and Southeast Asia, and that around 3000 BCE there was, for reasons unknown, a mass migration east by sea. The circumstances must have been desperate for such a dangerous action – essentially paddling into a vast ocean towards the rising sun – to be preferable to staying at home. These ocean-going emigrants settled various island chains – the Philippines by 1,500 BCE and Fiji by 500 BCE – and as local resources were exhausted and the population increased, groups would prefer, or would be obliged, to move on. The

Cook Islands were probably reached by 700 CE, then Rapa Nui (Easter Island) and Hawaii to the northeast by 900 CE. From the Cook Islands, bands travelled south in 1200 CE and reached Aotearoa (New Zealand), which was settled by waves of travellers during the following 100 years or so.

Maori traditions about arrival in Aotearoa are distinct. Specific locations where the ancestral canoes arrived are remembered, along with the identities and characters of the first ancestors. These all provided rich material from which to fashion distinct tribal identities. Myth and legend of course abound, but what are not recorded in these folk memories are precise dates of arrival or the more prosaic and practical details of travel, including techniques of navigation, which were evidently remarkable.

The first significant – if not the very first – encounter between Maoris and Europeans took place near Gisborne when Cook came ashore and, ominously, resulted in a Maori being killed. The Maori *iwi*, or tribe, established in the area appears to have been settled in what is now the small and remote fishing village of Whangara, a few miles up the coast from Gisborne. This village retains two particularly good examples of the architectural expression of Maori culture: the *wharenui*, or meeting hall, standing in its *marae*, or sacred ground. *Wharenui* are, in their purpose, repositories of tribal memories, pride, and identity, and were often constructed and decorated with great skill. One of Gisborne's *wharenui* was built in the 1890s, and the other in 1939. Both are fascinating.

Wharenui are sacred buildings. Their form represents the human body – the roof's ridge beam is the backbone, the roof timbers the ribs, and the walls represent the arms and legs. They are also the domain of the deified ancestors and protective spirits, usually represented in powerfully carved sculptural form. All that is important in the lifeof the tribe, and its individuals, happens in the *wharenui* and *marae*. They are where traditions are kept alive, where people come to sing, chant, and dance the *haka*, to discuss the future and the past, where they come when they are married, and where their bodies lie in state when they die. Each *wharenui* is dedicated to one main ancestor, usually seen as the founder of the tribe, and their image presides in glory over the top of the roof ridge above the main entry. Both of the Whangara *wharenui* are dedicated to Paikea, the whale rider, who is shown as a young tattooed man perched on the back of a sperm whale. According to legend, when Paikea's canoe sank while he was making the long journey to New Zealand, he was rescued by a whale and carried ashore.

Inside both *wharenui*, the story of the whale rider is told in a series of almost abstract paintings and carvings. But the smaller and older *wharenui* tells another story. At one end are images of the whale rider and Maori dignitaries, but at the other are paintings of the British royal family at the beginning of the twentieth century: King Edward VII, George V, and Queen Mary. The Maori evidently wanted to keep a foot in each camp. They wanted to make clear that they retained their own traditions but were also loyal members of the British Empire.

The *wharenui* are powerful and poignant. When they were built, the Maori were vulnerable. Now, following a resurgence in Maori culture, they are precious architectural monuments to a much-admired seafaring and warrior people.

38

STOCKHOLM PUBLIC LIBRARY
Stockholm, Sweden | 1922–33 | Erik Gunnar Asplund

The Stockholm Public Library, which was started in 1922 and not completed for over a decade, marks the period in the history of Western architecture during which classicism – in reinvigorated form – was re-establishing itself after the nineteenth-century dominance of the Gothic Revival. Other examples are the classical commercial architecture of Sir Edwin Lutyens, such as the Midland Bank Headquarters of 1924–39, in the City of London, and McKim, Mead, and White's glorious but now destroyed Roman-bath-like Pennsylvania Station in New York City, started in 1910. The library also marks what was, in many ways, a final flowering of classicism before Modernism overwhelmed it. The dethronement was not without its paradox, for Modernism, although antipathetic to the ornament and architectural hierarchies of classicism, was deeply dependent on classicism's systems of proportion and composition.

Even after the triumph of Modernism, from the 1950s until almost the end of the century, much classical architecture was built, but usually as a self-conscious riposte to the Modern. Thus after the Second World War classicism became a minority and reactionary architecture, while after the First World War it was – in the hands of a master like Asplund – very much a living thing, capable of inspired evolution and reinvention, in which sources were explored and exploited and forms chosen in response to function and landscape. In the process the innate compositional advantages of the classical tradition could be utilized to great effect, notably because classicism is capable of architectural rhetoric of varying degrees of intensity and because it represents an architectural language with a recognized meaning that can be comprehended. For example, traditionally in classical design an important public building proclaims its status in its urban or rural setting through the use of a portico, colonnades, or a large entrance door reached by a noble flight of steps, as in the Altes Museum in Berlin. This solution is as effective as it is obvious. Modernism, having rejected traditional forms and ornament, long struggled to find equally effective equivalents for expressing meaning or function.

At one level the Stockholm Public Library is a case study in the creation of a minimal architecture in which meaning is expressed in elegant manner. The architect's most obvious design strategy is to combine elemental forms – cubic podium, vast cylindrical drum, and hierarchy of square and double-square window proportions – that proclaim the classical roots of the building and suggest its importance. The design also, to a degree, suggests the building's function, with the drum as a direct expression of the library's reading room and book stacks. Asplund was an inspirational and experimental designer whose architecture developed from a rustic Arts and Crafts vernacular to an inventive and attenuated classicism that was a most individual expression of a style that became something of a fashion in Nordic countries in the inter-war years. From this, Asplund moved to a proto-Modernism – infused with the spirit of elemental classicism – which found sublime expression in his chapel, crematorium, and landscape at the Woodland Cemetery, Stockholm, completed at the time of his death in 1940.

The library, classical in its fundamental conception and details, also marks the transition in Asplund's architecture towards the simplicity and functionalism of Modernism. Asplund's early designs, produced from 1918 to 1920, were inspired by the solemn and simple late eighteenth-century French architecture of Étienne-Louis Boulée and Claude Nicolas Ledoux, which features large-scale primary geometric forms. The library, a fusion of drum and cube embellished with porticoes, was initially envisaged in heroic manner as a temple of learning on a low-rise acropolis. One particular building by Ledoux – the 1789 Barrière de la Villette in Paris, which is distinguished by a huge, flat-roofed drum rising above a lower, square-plan structure sprouting porticoes – provided the initial model for Asplund's design. But this evolved as Asplund gradually stripped away details so that, by the time construction started in 1924, the Stockholm Public Library largely proclaimed its classical allegiance through its system of proportions, with its windows no more than holes punched in flat walls colour-washed a strong, Mediterranean, orange ochre. Achieved through traditional pigments, such dramatic colour washes – warm ochres and blue and green verditer – reflected eighteenth-century Swedish traditions and became a distinguishing feature of early twentieth-century Nordic classicism.

The external details that Asplund retained are telling. The ground floor of the three-storey ranges wrapping around the drum, and which give the building a plan that is almost square, are rusticated and incorporate a finely detailed neo-classical frieze. But most importantly, the tall main door (modelled on the courtyard door of the Thorvaldsen Museum in Copenhagen) is approached by a stepped ramp framed by a classical architrave, proclaiming that this is a significant public building and making its main point of entry clear.

The door leads to an entrance vestibule that is dark, solemn, and with the character of an Egyptian tomb, its walls decorated, in low and subtle relief, with scenes from *The Iliad*. Stairs wind around the outside of the drum, serving galleries, while a straight stair rises into the drum itself. This, in contrast with the dark and constrained outer spaces, is lofty, spacious, and well illuminated, with light reflected down to the reading desks from its high white walls – the epitome of intellectual enlightenment.

39

FALLINGWATER (EDGAR J KAUFMANN HOUSE)
**Mill Run, Fayette County, Pennsylvania, USA | 1935 |
Frank Lloyd Wright**

As a personality, Frank Lloyd Wright can be provocative: too much the showman, the ostentatious dandy, even a poser; immersed in spurious mysticism and keen to peddle what he perceived as his almost talismanic Welsh origins. He was a man who borrowed easily from cultural history, natural history, and ethnic traditions – particularly those of Japan, when they were all the rage in the last decade of the nineteenth century – and who was forever evolving his architectural manner, perhaps even his beliefs and principles. This meant that around 1905 he was able to design in a sensitive Arts and Crafts manner –

conscious of context, materials, and tradition – but by 1956 he was happy to propose a mile-high skyscraper that confounded all issues of tradition and context. It was nothing more, of course, than a publicity-seeking fantasy.

Wright was, perhaps above all, a man wedded to self-promotion and immensely fond of the sound of his own voice, freely spouting his often wilful ideas, some borrowed, others a potent self-brewed concoction of the weird and the wonderful. He was, as one author has put it, 'a self-proclaimed and self-promoted genius for much of his professional life'.[8] But despite such failings – or perhaps even because of them – a man can be a great architect, can possess an inner talent that allows him to rise above the cacophony of his own ideas and survive the self-inflicted wounds of an emotive and chaotic life. Was Wright such a man? Certainly his career was far from conventional. By 1909, aged forty-two, it looked like it was all pretty much over for him. He had sparkled when young, had launched a prolific practice, and had even inspired a widely imitated style of domestic architecture with his distinctive contribution to the English Arts and Crafts-influenced 'Prairie School' houses that, with their broad, horizontal lines, deep-hipped roofs, and integration with the landscape, seemed to encapsulate the spirit of the prairies of the American Midwest.

But seemingly, early success had bored him. Wright abandoned conventional monogamous married life, and became embroiled in scandal. In 1914 at Taliesin (as his Wisconsin home and studio were called, after a sixth-century Welsh bard) he was tainted by a lurid multiple murder case, in which a mistress, Mamah Cheney, died along with her children and four others. This ghastly event unsurprisingly focused the public's attention on Wright's unorthodox life, but, learning no lessons, he soon became embroiled in an exhausting relationship with a new mistress named Miriam Noel and surrendered to the romance of spending a great deal of his time in Tokyo – during the decade after 1913 – overseeing the design and construction of his somewhat pedestrian Imperial Hotel. By the early 1920s Wright was largely forgotten, his Prairie-style houses long out of fashion as the new architectural forces of Modernism claimed American attention – and clients. The 1920s were a decade of professional failure and public humiliation for Wright, mostly the result of his own absurd actions: for example, the theft by his housekeeper of Noel's letters, which were sold to a Chicago newspaper and published to much public amusement and to Wright's mortification. But Wright fought back, regaining some reputation with his five Southern California Houses, designed during the 1920s, in which he rejected European historical models and instead fused a strange permutation of American indigenous and Japanese architecture. The houses were labelled as 'Aztec' by the popular press and were not altogether a success.

Wright struggled on – and then, in the mid-1930s, came Fallingwater in Pennsylvania, which proved a huge success. It turned the architect's failing career around, and he started his ascent to architectural legend and ultimately to sage and totem of the profession in the United States. It had been a close thing, and a keen example of the razor-thin boundary between the sublime and the ridiculous.

Fallingwater not only relaunched Wright's career but was, almost from the day of completion, an icon of Western domestic architecture, in which aspects of industrial Modern architecture of the 1930s were combined, in effortless and elegant ways, with some of the principles of the tradition-based Arts and Crafts movement. Among the key tenets of the Arts and Crafts movement was truth to site and materials. This meant placing a building to natural advantage and, as far as possible, wedding it to its site by using and displaying local building materials. On both these counts the house is a great success. It sits above a waterfall, offering spectacular views, and much of its exterior and interior structure is wrought of local stone. Indeed, rooting the house to the physical and mythical nature of the site includes the exposure, in front of the main hearth, of the natural stone that forms the peak of the rocky site and that, by tradition, anciently provided a place of meditation for local shamans. Other influences – notably from Japan and from the geometric orthogonal grid devised by Friedrich Froebel – are apparent in the overlapping and intersecting grid of squares and rectangles that determine the plan of the house and lead to a sequence of subtly related interior spaces, and in the glazed screens forming sections of external wall. Industrial Modernism is expressed most obviously by the cantilevered decks of reinforced concrete construction that evoke memories of the horizontal lines of the Prairie School. Predictably, the largest of these decks has failed due to cracking of the concrete and the penetration of water that has rusted and expanded the steel reinforcement.

But architecture – a thing of poetry and not just mere function – must not ultimately be judged by such technical failures. What man has the ingenuity to make, man also has the ingenuity to mend. And what of Fallingwater artistically? It must surely be admitted that some elements, such as the overtly vernacular masonry walls, are kitsch. But the house does possess an almost uncanny perfection in relation to its dramatic site, with vertical piers of masonry walling balanced brilliantly against the horizontal slabs of the concrete decks. Like its architect, Fallingwater is flawed – but also like its architect, it is clearly touched by genius.

40

CASA MALAPARTE

Punta Massullo, Capri, Italy | 1937 | Curzio Malaparte, with Adolfo Amitrano and Adalberto Libera

The Casa Malaparte is one of the oddest houses on earth. It was built by the journalist and writer Curzio Malaparte, who had been an enthusiastic Fascist during the movement's early days and took part in Mussolini's 1922 March on Rome that secured power for the dictator. In these days Malaparte was still beguiled by the romantic notion that Benito Mussolini was a long-awaited saviour of the nation, who would restore Italy's pride, economy, and world standing – and even rekindle the long-dead embers of Roman imperialism. By the early 1930s, however, the romance was over. Malaparte realized that Mussolini was an unprincipled adventurer, corrupt in his quest to gain and hold power, increasingly dependent on his overbearing German 'allies', and a totalitarian enemy of artistic and intellectual liberty. Malaparte expressed these views in a book, *Technique du Coup d'Etat*, published in 1931, and was consequently stripped of his membership of the Fascist Party and in 1933 condemned to internal 'exile' on the prison island of Lipari.

Malaparte was born Kurt Erich Suckert, but in 1925 – when in his late twenties and in his Fascist phase – changed his name to Malaparte. This was a pun on Napoleon's surname Bonaparte, which can be interpreted as meaning 'good side'; by contrast, Malaparte means bad or dark side. Lumbering himself with this somewhat laboured Surrealist joke-name no doubt reveals much about Suckert's sense of humour and perhaps his sartorial passion for the Fascists' black-shirt uniform. For most the joke would have worn thin as the hangover wore off, but Suckert stuck with his new persona – his *nom de guerre*.

In 1937 Malaparte was freed from Lipari due to the personal intervention of Mussolini's son-in-law and foreign minister, Galeazzo Ciano. Thinking it best to lie low, Malaparte retreated to what he described as 'the wildest, most solitary and romantic part of Capri where ... the island becomes ferocious', in order 'to build in the midst of ... nature'. And so, on a rugged rocky promontory overlooking the sea, Malaparte started to build his extraordinary house – part stylized Modernist fantasy, far removed from the stripped classicism that had become the house style of Fascist Italy, and part sculpture emulating the elemental force of nature.

The initial design was produced by architect Adalberto Libera, but soon he and his scheme were cast aside as Malaparte decided to take over the design of the house himself, with the help of a local Capri mason named Adolfo Amitrano. But work did not go smoothly, as Malaparte was dogged by vengeful Fascist authorities, who from time to time persuaded Mussolini to throw him back in jail. Consequently, the house was not completed until 1941.

Most refreshingly – and reassuringly, given the odd nature of its client and part designer – the Casa Malaparte is impossible to categorize. It is Modern in its flat-roofed form, stripped of significant historicist ornament, yet it scarcely seems to be a building. It is more a podium that Malaparte could mount to confront distant Fascist Italy, where independent thought had become a crime. As he famously said of his idiosyncratic creation, it is 'un casa come me' – a house like me. Presumably this refers to the defiance the house embodies and proclaims as its fights both the elements and artistic and political oppression.

The podium analogy is sustained by the house's most dramatic external feature – a grand flight of tapering steps rising from the rock on which the house sits to the flat roof that, sheer and frightening, hovers above the turbulent sea below. The only permanent feature on the roof is a curious curving and tapering screen that, if anything functional was intended, may have served as a windbreak for those basking on this exhilarating plateau. Or perhaps the screen is simply a brilliant artistic device to conceal the roof terrace, and its giddy-making unprotected edges, from those ascending the staircase. Distracted by the sculptural form of the screen, visitors will find that the exposed and unprotected nature of the terrace comes as quite a surprise when first encountered face to face.

The house's rendered stonework is painted a strong, red ochre and its plan is essentially a rather formal triple-square rectangle, so superficially it seems hardly calculated to fit harmoniously with the rugged and undulating forms of its clifftop site. And yet this is, in some strange way, what it does. Of course, it is a harmony achieved through contrast and perhaps can only be fully appreciated when arriving at the house by boat and scrambling up the steps hewn into the face of the rock on which the house stands. Seen from below, approaching at an angle, the house looks magnificent, ancient, and sublime; it adds to, and certainly does not diminish, the power of nature. The house has the feel of an antique temple rising from the sea, an impression reinforced by the sculptural form of the staircase, which evokes associations with the stepped crepidoma on which Grecian temples sit. I suppose this is what Malaparte intended.

The building's relationship with its dramatic site is reinforced by the interior, where windows are placed to offer the best view of coastline, rocky outcrops, and sea. The Casa Malaparte has a timeless quality and is profoundly thought-provoking. What more can be asked of architecture?

41

SÄYNÄTSALO TOWN HALL
Finland | 1948–52 | Alvar Aalto

The creation of public buildings using the architectural vocabulary of Modernism has proved a tremendous problem. Modernism is not an architecture of obvious rhetoric or of symbolism. To design a public building in the classical idiom, be it a museum or town hall, proved no problem to designers in the eighteenth or nineteenth centuries, for the language offered ready and appropriate metaphors. A colonnade or portico – perhaps pedimented – could proclaim a museum a temple of the arts (such as the British Museum in London or the Altes Museum in Berlin) or give a town hall or public building such as St George's Hall, Liverpool, the seal of ancient culture and the required air of civic or institutional dignity. Gothic architecture also offered designers forms and symbols that could express the elevated aspirations of public or grand commercial buildings, an excellent example being the Midland Grand Hotel, St Pancras, of 1865–76,by George Gilbert Scott. But twentieth-century Modernism is, in essence, a democratic architecture that by its nature eschews non-functional or structurally non-essential ornament and forms.

So it is fascinating to see how the language of Modernism was used by a master to realize a public building that – without resort to historicist imagery – evokes an aura of civic pride and dignity.

Alvar Aalto was born in Finland in 1898 and when young lived in a prosperous traditional rural community, next to a lake with a winding shoreline and surrounded by forest and undulating ridges. This gave him a strong sense of the beauty of nature and, more importantly, that nature and culture were complementary, one enhancing the other. This was a central idea that ultimately found powerful expression in Aalto's architecture – and nowhere more so, perhaps, that at Säynätsalo Town Hall.

Designed in 1948, the Town Hall was for a small community of only 3,300 people, and Aalto conceived this civic building as a town in miniature. He also chose a rising site – in a glade of trees that evokes the essence of Finnish landscape – that allowed him to give his miniature town hall the sense of an acropolis. And this was far from a purely romantic association given the ancient Greek origin of the democratic principles that Säynätsalo enshrines. On the central plateau of his acropolis Aalto placed a tiny town square. This square is reached by broad staircases, that on the west with grassy treads, which imbue the building with a strange sense of antiquity, of the elemental, suggesting that nature flows from the glade into the very heart of the building. This all triggers memories and associations and gives the building an emotional power way beyond that promised by its simple, functionalist elevations. These are wrought in humble brick, with virtually no ornament and with the only architectural flourish being an orchestrated irregularity of plane intended to give the elevations something of a natural-looking, cliff-like appearance. The architect's strategy to evoke nature and ancient civic architecture also defies scale – the building represents a big architectural idea enshrined in a minuscule body. You approach through the glade of trees and suddenly confront a staircase, tremendous in its ambition and through its association with the staircase leading to the propylaeum on the Acropolis in Athens, yet almost disarmingly small in scale. The same is true of the grassy square to which the stairs lead. Here, in the manner of a Renaissance campo or piazza, the more public parts of the building – library, reading rooms – are grouped around its periphery. This was clearly intended as the 'cultural' heart of the community, a place for events and leisure activities.

The interiors of the Town Hall express the same austere brick aesthetic as the exterior elevations. All is simplicity, with even the council chamber being little more than a double-height brick box – admittedly placed on the highest level of the site to, presumably, represent its 'elevated' function. Only in the roof structure inside the council chamber is there some sense of ornament, with the timber-clad sloping ceiling supported on struts fanning out from the centre of tie beams. But this flourish is really no more than an honest expression (in the Gothic spirit) of the materials and methods of construction used, and an acknowledgement of good carpentry skills.

This building not only demonstrates a solution to the problem of designing a 'monumental' public building in Modernist vocabulary but is also a memorable expression of 'democratic' architecture. Here the architect gives meaning to his design by evoking the essence of antiquity while organizing it around the needs of the people of the community. The building is, in essence, a celebration of the power of history and the virtues of democracy.

42

NATIONAL PARLIAMENT
Sher-e-Bangla Nagar, Dhaka, Bangladesh | 1962–83 | Louis Kahn

Louis Kahn is one of the most enigmatic architects of the second half of the twentieth century. His mature architecture, in its fusion of historic inspiration and Modernist functionalist principles – coupled with Kahn's intuitive genius for form – is among the most successful of its age. And among the best of Kahn's late work, in its poetic power and geometric purity, is the National Parliament at Dhaka.

Kahn was born in Tsarist Russia in 1901 and emigrated with his family to the United States at the age of five. The family was poor but Kahn's natural talents gained him admission to the University of Philadelphia, where he completed his architectural education in 1924. He had a job in the office of the Philadelphia city architect and in 1928 was able to make his first architectural tour of Europe, where the mass and solidity of medieval military and cathedral architecture fired his imagination. The impressions gleaned during this tour were reinforced in the early 1950s, when Kahn was architect in residence at the American Academy in Rome. During this time he travelled through Italy, Greece, and Egypt, absorbing the power – the volumes, the massive construction and scale – of ancient buildings, structures that were seemingly conceived to outlast time itself. The experience was transformative.

For Kahn, history became a living thing. He resolved that his architecture should connect with – should echo – the timeless virtues of these venerable and mighty works. Space and light could bring a building to life and make it a suitable domain for the human spirit. As he said, 'all material in nature' is perceived 'as light … so light is really the source of all being'.[9]

Kahn also studied, as if from first principles, the nature of ancient construction, and realized that traditional materials, used well, can give a building beauty and dignity. As Kahn put it, the brick is 'always talking to me'. He once described such a conversation: 'You say to a brick, "What do you want, brick?" And brick says to you, "I like an arch." And you say to brick, "Look, I want one, too, but arches are expensive and I can use a concrete lintel." And then you say: "What do you think of that, brick?" Brick says: "I like an arch".[10]

Kahn revelled in the visceral excitement of structure being exposed and legible, as with the vast brick relieving arches revealed in the massive walls of Trajan's Market and the Pantheon in Rome. The poetry of primal geometry and the elemental forms of such ancient structures amazed him. For Kahn, as with the eighteenth century French architect and theorist Étienne-Louis Boulée, abstract, geometric forms, simplicity, and the ordered manipulation of scale and space were the basis of beauty – a beauty that is not relative or based on cultural associations but is absolute, timeless, and universal. Of Boulée, and his French near-contemporary Claude-Nicolas Ledoux, Kahn observed that their visionary projects display an 'enormous desire … to express the inspirational motivations of architecture'; these buildings 'were not projected to satisfy function … but belonged to the challenge against narrow limits' and possessed a 'stupendous … audacity belonging to architecture'. In short, he said, 'Boulée is, Ledoux is, thus architecture is'.[11]

Many of these observations ring true when applied to Kahn's own work – especially his great final project, the National Parliament for the new nation of Bangladesh. Kahn was first approached in 1962, before Bangladesh existed, by the authorities in Pakistan to design government buildings for Dhaka, the provincial capital of East Pakistan. During the slow and painful design process the nature of the project changed fundamentally when, following civil war, East Pakistan became in 1971 the independent state of Bangladesh, with Dhaka its capital and Kahn's buildings housing its sovereign parliament. So Kahn was in at the birth of a nation and had to conceive the form of a place of assembly for the new government and express through his architecture national pride and identity.

The buildings do not fail in this task. Kahn's quest for essential beauty through primal geometry, the truthful expression of material and means of construction, the combination of rugged in-situ cast concrete and local brick, and the creation of interiors infused with natural light gives the parliament complex a magic quality, a sense of solidity and of dignity. And, in homage to the water-world of Bangladesh, the buildings rise from lakes, in which their forms are echoed and animated by the ripples of the water. Here are memories of castles, of the fortified French cathedral of Albi with its cylindrical brick piers, of Rome, of Ledoux and his passion for the circle and the sphere – and all are fused brilliantly to create an individual and coherent architecture.

43

TRELLICK TOWER
Golborne Road, London | 1967–72 | Ernö Goldfinger

In the early 1980s in London, Ernö Goldfinger was a figure to be feared. He was one the great beasts of the British architectural profession, with a most alarming reputation. His architecture revealed that he was determined and uncompromising and that he had no time for courting popularity, following fashions, or seeking to please. In fact, he seemed to have no interest in giving the public and his clients – in the latter part of his career almost exclusively the politicians and professionals of the Greater London Council – what they might think they wanted. This, from the 1970s onwards, tended to be low-rise housing of more or less traditional form. For Goldfinger this was retrograde and a betrayal of logic and principle. He remained determined to give his clients what he believed was good for them – or more particularly, to give them architecture that was bold and powerful and that also provided decent homes and environments. At the time Goldfinger proclaimed that 'the client does not care about architecture, he cares whether the building works. The architecture is my own private pleasure.' And Goldfinger's pleasure was not to bow to the prevailing tastes of the time – which he evidently believed were timid and reactionary – but to stick to his guns and provide buildings true to the vision of the heady days of the Modern movement, when the duty of an architect was the welfare of the people. For Goldfinger this duty embraced high-rise housing, inspired by such prototypes as the Narkomfin Building in Moscow and Le Corbusier's Unité d'Habitation in Marseille that offered the possibility of healthy and spacious homes with pleasing plans, flooded with light and blessed with splendid prospects.

Goldfinger's last great project – Trellick Tower in Westbourne Park – is a thirty-one-storey slab block, completed in 1972, that was notable (indeed, for many in the 80s notorious) for its expression of Corbusian theories about high-rise living in the city, when these theories were deeply unpopular and were being blamed for the destruction of traditional urban communities and for the introduction of any number of social evils. Trellick Tower, together with its related low-rise architecture, was a slightly taller and reconfigured development of Goldfinger's earlier Balfron Tower in Poplar, east London, of 1963–67, commissioned by the London County Council.

In the early 1970s I got to know both these towers well. As a child of the time I naturally disapproved of them, and the role I imagined they had played in the sweeping away of streets of terrace houses and in the disruption of established communities. But, as I had finally to admit to myself, both towers possess a fatal attraction, a sublime and unsettling architectural presence. They seemed to cast a spell woven out of the fabric of their construction, their scale, their arrogant proclamation of moral and ethical certainties that others doubted. They were defiant and audacious, and possessed a weird, almost guilt-inducing visual poetry. To enjoy them was almost, for my generation, to commit an urban sin. The most thrilling moment, I remember, was to see the newly completed Trellick Tower from the newly completed Westway. It was a ballet performance. Speeding along Westway, the observer would first see the tower from the diagonal, distant but intriguing; then it would rapidly grow to a mighty slab, and suddenly a vertical sliver of light would appear in its bulk as the lift shaft detached itself visually from the bulk of the building. Finally, as the road curved, the tower would shrink to a mere sliver when only its narrow width was visible. This was clearly great and beguiling architectural theatre.

I eventually met Goldfinger face to face in the early 1980s. I arrived at his house in Willow Road, Hampstead – which he had designed in 1938 – keyed up and ready for combat. I was to interview him for an article in the *Architects' Journal*. I knew his reputation, that he did not suffer fools gladly, that he was impatient, and I imagined this would be a confrontation, a power struggle. Well, I was a fool. Ernö was charming, gracious, a wonderful and witty host and storyteller – a gentleman of the Austro-Hungarian empire who had been born in Budapest and had reached maturity in Paris of the 1920s, and who had been intimate with many of the great intellects and artists of interwar Europe. We soon found much in common besides architecture – I seem to remember military history, nineteenth-century uniforms, and London were favoured subjects. Ernö mixed a mean cocktail, and his wife Ursula was as quiet and self-effacing as Ernö was boisterous and self-centred. But although this was the real Ernö, it was of course only one part of the man. He had to have a dark and obsessive side, or his architecture would not be what it is.

The full complexity of Goldfinger's character is revealed by architect James Dunnett, who worked for Goldfinger from the 1970s. Over the years Dunnett has written beautifully, and with much insight, about his temperamental and often bullying 'Master'. Rarely has an architect been so well – and perceptively – served by a biographer. Dunnett writes of the 'terrible beauty' of Balfron and Trellick towers, their 'air of menace', and the way in which Goldfinger's use of raw and bush-hammered concrete has rarely made the material seem 'so beautiful'. Dunnett also observes that Goldfinger chose as a source of inspiration 'the artifacts of war', with the sheer concrete walls of the lift shafts at both towers 'pierced only by slits, cascading down the façade like rain' to 'impart a delicate sense of terror', and upper storeys like the bridges of warships. This, states Dunnett with a tremor of horror, 'is Stalin's architecture as it should have been' – and yet, 'under a low evening sun one has the feeling of participating in an heroic landscape'.[12]

Each must come to their own view about Goldfinger's architecture. Trellick and Balfron towers are now listed buildings, so – from the official point of view at least – although strangely anachronistic, these key Goldfinger creations have stood the test of time.

CHAPTER 3

SACRED

HUMANS BUILD FOR VARIED reasons: to create shelter and communities, for security, to express worldly power, for sensual pleasure and for the delights of visual beauty. But some of the greatest architecture created through the millennia has been made to honour and commemorate the gods, to evoke images of paradise on earth and give tangible expression to the sacred, to mark the mystery of death and to embody speculations about life beyond death. In many ages, for example in ancient Egypt, in medieval Christian Europe, in the Muslim Middle East, and in Hindu India, sacred art and architecture have been the primary works of man – the most significant, the most valued, and the most sublime.

The quest by various cultures to give expression to the sacred, to create heaven or the universe in microcosm and in accordance with perceptions of natural and divine laws, to offer spiritual guidance, inspiration, and assurance, has arguably produced some of the world's greatest architecture.

The wonders of sacred architecture include much of the surviving architecture of ancient Egypt, such as tombs and mortuary temples like that of Hatshepsut, built around 3,450 years ago, one of the oldest, most sophisticated, and most awe-inspiring pieces of architecture in the world. In addition there are the sixth-century Hagia Sophia in Istanbul and the seventh-century Dome of the Rock in Jerusalem, which are among the earliest monuments to Christianity and Islam respectively and are, in turn, masterpieces of bold engineered structure and complex, subtle, and coded sacred design. The ninth-century Malwiya at Samarra, Iraq, is one of the world's first minarets and remains one of the most stunning abstract pieces of large-scale sculptural architecture. Durham Cathedral in England and Chartres Cathedral in France are among the most impressive and enigmatic architectural and engineering achievements of medieval Europe; the late-twelfth-century Minaret of Jam in Afghanistan is one of the most beautiful – and remote – Islamic monuments of Central Asia, while the Buddhist Bayon in Angkor Thom in Cambodia – of the same century – is a haunting masterpiece of sacred architecture in Southeast Asia. In Africa, the early thirteenth-century rock-cut churches of Lalibela in Ethiopia and the sun-dried brick and mud-built Great Mosque at Djenné are idiosyncratic and very distinctive sacred structures. In India, a sub-continent blessed with a wealth of architecture for all the world's great religions, the mandala-like temple town of Sri Ranganathaswamy Temple at Srirangam represents the complex world of Hindu sacred architecture, while the seventeenth-century Taj Mahal at Agra represents Islamic achievement. Late nineteenth- and twentieth-century European Christian architecture is represented by two buildings that can be seen as diametric opposites. The Temple Expiatori de la Sagrada Família in Barcelona has been under constriction since 1883 to the neo-Gothic-cum-organic Art Nouveau designs of Christian visionary Antoni Gaudí, while the Modernist Notre Dame du Haut at Ronchamp, France, was designed by Le Corbusier – a man of no conventional religious faith, perhaps even an atheist.

44

MORTUARY TEMPLE OF HATSHEPSUT
Deir el-Bahri, Luxor, Egypt | c.1450 BCE | Senenmut

The West Bank of the Nile at Luxor (anciently Thebes) is the land of the dead, the place where the sun sets – where it dies – each night. The East Bank, where the modern city nestles, is the land of the living, where the sun is reborn each morning, and life proclaimed.

The ancient structures on the West Bank are mostly related to death, internment of the body, and the trials and tribulations of the afterlife. Most notable are the mortuary temples – buildings intended to keep the names and memories of their builders alive long after the death of the body. These structures, referred to as 'Houses of the Millions of Years', were to last for eternity; they were attempts to achieve immortality through architecture, machines of rebirth intended to sustain the soul in the underworld.

As well as the ruins of mortuary temples, the East Bank contains tombs in their thousands upon thousands; tombs of the great and mighty, tombs of the rich and influential, and even humble tombs of the more common people. The tomb and mortuary temple of one of the most intriguing and mysterious characters in Egyptian history belong to a woman who ruled Egypt nearly 3,500 years ago, but as a man, as a pharaoh. This ruler's throne name was Maat-Ka-Re, but she is best known to history by one of her other names: Hatshepsut.

Her tomb is little known, and never open to the public – it's too ruinous, remote, steep in ascent, and barren to make that feasible or generally profitable. But I gained consent and one bright summer morning descended into tomb KV 20. This, as far as anyone can be certain, is where Hatshepsut was originally interred. The tomb was evidently never completed – the walls are rough-hewn, with no sign of paint or plaster, the floor is covered with a thick layer of dust and rubble, and the descent is abrupt and treacherous.

The first burial chamber lies about 200 metres (650 ft) below the entrance. Beyond, there is another, far smaller one. This is where Hatshepsut's sarcophagus was discovered in 1902, together with that of her father, Tuthmosis I. It seems she had wanted both their bodies to rest together for eternity. When found, the sarcophagi were empty, and the only hard evidence that Hatshepsut had been here was a stone vessel inscribed with her name. There is one curious fact about this unusually deep tomb and its out-of-the-way location. The tomb builders, as they wound deeper underground, had been looking for something: for a spot immediately beneath a structure some hundreds of metres overhead. Hatshepsut's burial chamber is immediately below the inner portion of the Mortuary Temple that she constructed in the bright sun above. Both tomb and temple were evidently intended to work together, to ensure the survival and safe journey of Hatshepsut's soul after death and to grant her spirit immortality.

The Mortuary Temple had one primary purpose – to preserve the memory of Hatshepsut's name after her death and so help ensure her perpetual life. Egyptians believed that as long as a name lived so did the soul, and if a name were destroyed or forgotten then all would be lost. And to live after death, the soul – made up of five parts, including the *ka*, or spirit and life force – had to be sustained through prayer and offerings. So the architecture of mortuary temples is indeed the architecture of eternity.

The temple, now substantially restored and reconstructed, consists of a large forecourt, almost square in plan, that was originally entered via a pylon, long since gone. Once, a processional causeway lined with sphinxes led from the pylon straight to the distant Nile. At the west end of this first forecourt is a row of piers, of rectangular section, that shelter a gallery containing fragments of wall paintings and hieroglyphs. Set between these piers is a wide ramp that rises up to a second court. This courtyard is also roughly square, with the shorter side set parallel with the court below. Along its west side is another row of piers, again sheltering a gallery. On the north side of this court is a row of simple fluted columns anticipating Greek Doric columns by nearly 1,000 years. At the north and south side of this gallery are two structures that, although skilfully integrated into the architecture of the court, are also distinctly different. The structure to the south is a temple to Hathor, the nurturing goddess and emblem of female power who evidently had particular meaning for Hatshepsut.

The structure to the north of the gallery is a shrine to Anubis, the jackal-headed god of mummification, and its well-preserved ceiling recalls one of the key characteristics of Egyptian temples. It portrays the firmament, which was the starry home of the gods – and since gods were cosmic beings, temples in their honour were conceived as the world in miniature.

From this second court another ramp rises to a narrow terrace that once contained many statues of Hatshepsut in the guise of Osiris, the god of the underworld, the lord of the dead – and thus she was portrayed as both pharaoh and god. West of this terrace is a third court, roughly Golden Section in proportion (the ancient ideal of natural beauty, approximately 1:1.618 in plan). Beyond this court is the temple's Holy of Holies – 'the sublime of the sublime' – and this was a very spiritually charged place indeed, because it was a shrine to the great God of Thebes, Amun-Re, whom Hatshepsut claimed as her true father.

Extensive twentieth-century restoration has revealed the architectural power of Hatshepsut's temple and confirmed it as one of the most mesmerizing monuments of ancient Egypt.

45

TEMPLE OF SOLOMON
Jerusalem | c.1000 BCE

The Temple of Solomon now lives largely in the imagination, but it is perhaps the most influential structure ever built by man. Initially constructed perhaps 3,000 years ago, it has had a profound influence on Jews, Christians, and Muslims because, despite being built by man, those of religion believe it was designed by God – that it was God's house on earth within which were enshrined secrets about creation, about the future and the past, that it embodied laws about the design of sacred buildings, that it established the principles of architectural beauty and was responsible for the origin of classical architecture. It was also the initial home of the Ark of the Covenant, God's throne on earth.

The only information we now have about the Temple – and there were evidently a series of structures on the same site, since the Temple was regularly rebuilt after a sequence of calamities – is preserved in various books of the Old Testament and in fragments that survive on the fabled Temple Mount in Jerusalem. The existing physical remains relate to the Temple as restored, enlarged, or rebuilt by King Herod I in about 20 BCE, and it was this version of the Temple that the Romans largely destroyed in 70 CE, following a Jewish uprising. The destruction of the Temple was accompanied by the exile of the Jews from their homeland and the beginning of a nearly 2,000-year-long diaspora. It is the remains of Herod's Temple, notably the retaining wall around the Temple Mount – including the West or Wailing Wall – that survives today.

Biblical accounts suggest that there were three material Temples in Jerusalem and one visionary one. The first Temple, built by King Solomon around 950 BCE following instructions from his father King David, is recorded in separate Biblical accounts. Significantly, this Temple, although divine in inspiration, was the handiwork of man. According to 1 Kings 7: 13–20, Solomon commissioned construction from Hiram 'out of Tyre ... a widow's son ... filled with wisdom and understanding and cunning to work all works'.

The measurements of this first Temple are described in great detail in I Chronicles 28, II Chronicles 3: 1–17 and 4, and I Kings 6: 1–35 and 7: 13–20. All suggest an orthogonal, rectangular, or cubical temple with, for example, the 'house of God' in the Temple being a triple cube in proportion (II Chronicles 3: 3). Key details are also described. For example, 'David gave to Solomon his son the pattern of the porch ... and of the place of the mercy seat ... and he reared up the pillars before the temple, one on the right hand, and the other on the left, and called the name of that on the right hand Jachin, and the name of that on the left Boaz' (I Chronicles 28: 11 and 2 Chronicles 3: 17). According to legend, these two pillars – marking the route to the Holy of Holies – were hollow and were the means by which secret knowledge was saved from destruction by fire and water. The story is told in the Apocrypha and could be of Babylonian origin.[1]

It appears that the first Temple was attacked soon after completion. Sheshonq I, the first king of the 22nd Egyptian dynasty, who reigned *c.* 943–922 BCE, conducted at least one invasion of Palestine, in the course of which he raided Solomon's Temple and carried off its treasure, 'which Solomon had made'.[2]

According to the Bible, the Lord spared Jerusalem from complete destruction, but in around 586 BCE the Temple was attacked again and this time destroyed. The enemy was King Nubuchadnezzar, who carried the Israelites – along with the Temple's treasure – into exile in Babylon. It is possible that the Ark of the Covenant was stolen, hidden, lost, or destroyed at this time – unless Menelik, Solomon's son by the Queen of Sheba, had already carried it to Ethiopia, where Ethiopian Orthodox Christians believe it remains.

The construction of the second Temple was orchestrated in about 520 BCE by Zorobabel, a Persian official of Jewish descent. Zorobabel used his diplomatic skills most successfully on Cyrus, the King of the Persians, soon after the release of the Jews from Babylonian captivity. The circumstances surrounding the construction of the second Temple are described in 1 Esdras, chapter 2: 'The Lord raised up the spirit of Cyrus the king of the Persians, and he made proclamation … saying … the Lord of Israel, the most high Lord, hath made me king of the whole world. And commanded me to build him an house at Jerusalem in Jewry.'

The key measurements of the second Temple are contained in Ezra 6: 3–4: 'Let the … height thereof [be] threescore cubits, and the breadth therefore threescore cubits. With three rows of great stones, and a row of new timber: and let the expenses be given out of the king's house.' So it had key dimensions of 60 cubits, and sounds very much like a cube in form – although an equality of height and breadth could also be achieved by the creation of a cylinder with equal height and diameter. It was this second Temple that was restored, enlarged, or rebuilt by Herod I in about 20 BCE and almost completely destroyed by the Romans in 70 CE.

The visionary Temple is described in Ezekiel chapters 40 and 41, in verses of geometrically inspired ecstatic prophesy: 'So he measured the court, an hundred cubits long, and an hundred cubits broad, foursquare' (40: 47); 'Afterward he brought me to the temple, and measured the posts, six cubits broad on the one side, and six cubits broad on the other, which was the breadth of the tabernacle' (41: 1). The court in front of the temple's 'most holy place' measured twenty cubits square (41: 4). The description is detailed, complex, and confusing, but essentially the Temple was formed with cubical structures and square courts.

It must have been these images that inspired St John the Divine when, in chapter 21 of the Book of Revelation, he described the 'New Jerusalem', the celestial city: 'And I ... saw the holy city, new Jerusalem, coming down from God out of heaven ... and [it] had twelve gates ... and the wall of the city had twelve foundations ... and the City lieth foursquare and ... the length, and the breadth, and the height of it are equal.' In his vision John saw an angel measure the walls of the New Jerusalem, which were 'an hundred and forty and four cubits'.

The New Jerusalem could be interpreted as cubical or cylindrical, but it seems certain that these Biblical descriptions of the Temple and of the New Jerusalem have inspired the designers of synagogues, churches, and mosques over the centuries, each attempting to reproduce the sacred plans and proportions of Biblical sacred architecture. And from the Renaissance onwards there were numerous speculative reconstructions of the Temple, notably that of the Spanish Jesuit Juan Bautista Villalpando, published in 1596, which was based largely on Ezekiel's description. Villalpando argued that classical architecture emerged with the Temple and so is, in origin, the creation of the God of the Jews and of the Christians, and not pagan.

46

ST CATHERINE'S MONASTERY
Sinai, Egypt | Early 6th century CE

The Monastery of St Catherine, in the Sinai, Egypt, is one of the world's most evocative and sacred sites for Christians and for Jews. The monastic buildings, looking much like a small, walled medieval town, quaintly square in plan, stand below a mighty peak long ago identified as the Biblical Mount Sinai and on the site where, it is said, Moses had his first, awe-inspiring encounter with God. After fleeing Egypt, where he had murdered an Egyptian who had been mistreating Israelite workers, Moses took refuge in the Sinai Desert. There he married a young shepherdess, a Bedouin, and while looking after the flock of her father, Jethro, he had a transformative experience that ultimately led him to liberate the Israelites in Egypt. He saw a bush that burned with fire but was not consumed, and, as explained in Exodus 3, 'God called unto him out of the midst of the bush, and said, Moses, Moses. And he said, Here am I.' Moses moved towards the flames and the voice, but was stopped. God said, 'Draw not nigh hither; put off thy shoes from thy feet, for the place whereon thou standest is holy ground.'

The monastery stands on that holy ground, the place where God walked on earth, because it has been claimed, for the last 1,800 years at least, that it occupies the site of the burning bush.

There has been a Christian monastery on this site since the late third century CE, but in its existing form the monastery dates from the early sixth century. At that time this was a largely Christian land – it was a hundred years before the rise of Islam – and the Sinai stood on the edge of the mighty Byzantine Empire, then the greatest Christian power in the world, centred on Constantinople (modern Istanbul). In 527 CE Emperor Justinian decided that the collection of humble buildings gathered around the site of the burning bush, and below the mountain where Moses received the Ten Commandments, should be rebuilt in grand style and protected from marauding pagan Arabs by being placed within a stout wall. It seems the emperor, to honour his wife Theodora, wanted to show his respect for this sacred site and leave his mark on this holy land.

So St Catherine's was born in its present form, an architecturally admirable monastery that would serve also as a frontier fortress. Its reconstruction became an imperial project, with the master builder Stephanos of Aila and workmen organized from, and answerable to, Constantinople.

All the monastery's main buildings are packed within the tall, towered wall, with only a garden, its great Cyprus trees looking like a corner of paradise beleaguered in the desert, lying exposed and undefended.

The original large door to the monastery was blocked long ago, and until relatively recently the only way inside – for people and goods – was by being hoisted up to a high-level hatch. Now, however, there is a route by means of a small door in the western wall. This leads, via narrowed and angled passageways with stout inner doors, to a small and well-defended entrance court just inside the wall. It is evident the monastery was well and scientifically fortified and a most serious frontier fortress.

The key building is, predictably, the church. It was built around 550 CE, and it enshrines the secret meaning of St Catherine's. By the west door is an inscription from the Psalms: 'This gate of the Lord, into which the righteous shall enter' (Psalm 118: 20). This suggests that the church was a place open only to initiates, to the baptized. The sixth-century doors are loaded with delicate carvings of creatures cavorting among flowers and plants, and the ceiling beams high above, just as old, show similar scenes – images of paradise. The beams also bear the names of Emperor Justinian, of his wife, and of the master builder, Stephanos. This was, when new, clearly a very important place, and its meaning is expressed in the details and form of its architecture.

The church is a basilica in plan, with the nave separated from the aisles by mighty monolithic granite columns. The size and immense weight of these columns (each now unfortunately clad with plaster) reveal what a serious construction project this was.

There are twelve columns, six on each side. Twelve is a recurring number in the design of the monastery – for example, it contains twelve chapels – and is a number that possesses much ancient meaning: there are twelve months of the year, twelve apostles, and twelve signs of the zodiac. One of the chapels offers another meaning, however; it is dedicated to St John the Divine, the author of the last book of the New Testament, the Book of Revelation, which was written at about the time this monastery was founded. It's a prophetic text that unveils the way in which this world will end and how the new holy world of Christ will come about. It also describes the cubical form of the 'New Jerusalem', with its twelve gates, twelve foundations, and wall 144 – or twelve times twelve – cubits long.

The twelve chapels of St Catherine's, and the twelve columns in the church, must represent the 'twelve foundations' of the Holy City, and the length of the monastery's walls is more or less 144 cubits. So the Bible seems to have acted as a design guide for the construction of this miniature sacred city, built in similitude of the New Jerusalem.

The importance of the Book of Revelation in the conception of the monastery is confirmed by one of the strange and varied column capitals in the church. It shows a cross from which are suspended the Greek letters Alpha and Omega, which are a direct reference to the Book of Revelation, in which Christ states: 'I am Alpha and Omega, the beginning and the ending ... which is, and which was, and which is to come.' So surely this monastery was intended as a prophecy, as a promise of the world to come – it is the word of God in stone.

47

HAGIA SOPHIA

Istanbul, Turkey | 532–37 CE | Anthemius of Tralles, Isidorus of Miletus

Hagia Sophia, a name that means 'holy wisdom', was commissioned by Emperor Justinian I in 532 CE and, when completed five years later, was regarded not only as the greatest church in the eastern Roman Empire – Byzantium – but as the greatest church in the world. In both scale and architectural ingenuity, it outstripped the Holy Sepulchre in Jerusalem and St Peter's Basilica in Rome. Like the Anastasis of the Holy Sepulchre, it is circular and domed, but of such a scale and elegance of construction that it appeared to defy the laws of nature. The architects of Hagia Sophia – Anthemius of Tralles and Isidorus of Miletus – seem to have been determined to eclipse the structure that must have been their direct inspiration, the domed and massively constructed Pantheon in Rome, which was completed in about 130 CE as a pagan temple to all the gods. The span of the concrete dome of the Pantheon is nearly 12 metres (40 ft) greater than the 32-metre (104-ft) diameter dome of Hagia Sophia, but the height to the top of the Hagia Sophia dome is, at 55 metres (180 ft), significantly greater than the corresponding dimension in the Pantheon. More importantly, Hagia Sophia achieves its stupendous scale with an elegance, refined minimalism, and scientific ingenuity that puts the Pantheon in the shade. Indeed, the lofty dome of Hagia Sophia seemed, when first completed, to be an almost supernatural incarnation of God's creation, a celestial hemisphere that hovered, virtually unsupported, between heaven and earth. It was pioneering – nothing quite like it had ever been created before. Justinian, when he saw the completed interior for the first time in 537, revealed another inspiration for the building: 'Solomon', he exclaimed, 'I have surpassed thee.' Justinian and his architects must have been trying to recreate, in the new and great city of Constantinople, an epic vision of the Temple of Solomon, the divine building executed to the instructions of God. There are many lengthy and complex Biblical descriptions of the Temple, but one (Ezra 6: 3–4) suggests simply that its breadth and height were equal – a proportion achieved with a cube or with a cylinder of equal height and diameter. These primary forms are united in Hagia Sophia's cubic and domed central volume.

When completed, Hagia Sophia proved to be one of the great inspirations for the design of sacred buildings in the Middle East, North Africa, and Europe (the old domain of the Roman Empire), be they Christian or Islamic. Hagia Sophia now looks like a sacred mountain of a building, vast and elemental. The huge dome, with its dark covering, is now framed by much later minarets added in the late fifteenth and sixteenth centuries, when the church was converted to a mosque after the Turks seized Constantinople in 1453. The exterior is complex, with the portions added over the centuries in response to earthquake damage or changing use, now confusing the fundamental structure. But what is clear is that the building is really all to do with its interior space – and, of course, the most important part of the interior is the celestial dome, a homage to the wonder of God's creation. Much of the exterior – the various buttresses and terrace-like blocks of structure – exist solely to keep the great dome in place.

The building was entered through a door in the west that led to a now long destroyed cloister-like atrium. From here, the existing entrance hall was reached. This route passes beneath a wonderful early mosaic that reveals the source of the holy wisdom after which the church is named. In the centre of the mosaic is the Virgin Mary, flanked, and being bowed to, by Emperor Constantine, who holds a model of his city of Constantinople, and by Justinian, who holds a model of Hagia Sophia. They are dedicating their creations to the Virgin, asking her protection. Beyond the mural is the narthex and then the main, curved, interior space. To enter this space is to enter an ideal world, over which the dome of heaven floats. It represents a magical harmony between the square and the circle, the time-honoured geometrical expressions of the material of the sacred worlds. Down below, the realm of man, all is right-angular. Above, all is curved – the realm of God.

The way in which the rectilinear basilica-like lower portion of the interior transforms, at it rises, into a series of domes and semi-domes, is miraculous. First, the four main piers of the basilica sprout four huge semicircular arches that leap from one pier to the other. These four arches are linked by four forms called pendentives – essentially concave triangles – that join to form the circular ring on which the base of the drum sits.

This is what you see when standing in the nave. But this is only the obvious part of the structural system, and because this structure appears so minimal and elegant, the dome appears to float – especially since its lower portion is pierced by a closely set row of windows, making the junction between dome and supports look transparent, almost non-existent. This apparent lightness is the genius of the design, for it is achieved by sleight of hand, by clever structural tricks, and these tricks work because they are grounded on a thorough understanding of the science of construction, on a wily ability to hide or disguise structurally essential but bulky bits of the building, and not least, on a brave determination to take chances. This was an experimental structure, and its designers never lost their nerve. They pushed what was possible to the edge – and perhaps beyond the edge in the circumstances, because part of the central dome collapsed less than twenty years after completion due to the entirely predictable effects of an earthquake.

But this was no ordinary building – it was the great house of God, and the response to the disaster is revealing. Rather than showing caution during reconstruction, the dome was heightened, although with the sensible addition of a row of squat external buttresses. The real structural brilliance comes with the elegant way in which the central dome is handled. Its outward, lateral thrust is largely balanced by the opposite and roughly equal counter-thrusts exerted by two half-domes opening to the east and west and by the small domes at the four corners of the nave. Sturdy buttresses to north and south transfer much of the horizontal thrust of the dome into the rest of the structure and carry it down to the ground. All is in equilibrium, is poised; the laws of science, of nature, are used to keep the structure standing.

The building is a prayer in brick and stone rising up to heaven – a vast structure that, paradoxically perhaps, achieves spiritual power through the supreme understanding of the material world.

48

DOME OF THE ROCK
Jerusalem | 689–91 CE

The Temple Mount in Jerusalem contains physical remains and evocations of the long-lost Temple of Solomon buildings. There are King Herod's walls of c.20 BCE, and the al-Aqsa mosque, the third-holiest site in Islam, which dates in part from the eighth and eleventh centuries and from soon after the Crusaders took Jerusalem in 1099, when it was transformed by the Knights Templar into their headquarters. But there is also the Dome of the Rock, which is one of the most fascinating, beautiful, and strange buildings on earth.

The Dome was perhaps built as a Muslim riposte to the domed church of the Holy Sepulchre. Dating from the fourth century, the Holy Sepulchre was for centuries the most important Christian church in the world, and for some denominations is still the centre of the Christian universe.

After the Muslim conquest of Jerusalem in 637 CE it became increasingly inappropriate that the most impressive sacred building in the city should be Christian, so finally, in 689, the Umayyad Caliph Abd al-Malik ibn Marwan had the Dome of the Rock built so that the Holy Sepulchre would no longer 'dazzle the minds of the Muslims'.

But although the origin of the Dome lies in a worldly gesture of triumph, its architecture was – and remains – serene and proclaims the importance of peace and virtuous conduct.

The Dome of the Rock is not a mosque, it's a shrine – but it's not even that, really. It's more a three-dimensional calendar and a statement about the journey from this world to the next. In its geometry, it is a sophisticated interpretation of the cubic form of the lost Temple while also referring to the Byzantine architecture of the Holy Sepulchre.

The design of the building is based on the square and its permutations, notably the circle, which is a form that fits perfectly within the square or is defined by its four corners. The square and the circle – and their three-dimensional equivalents, the cube and the sphere – are the basic building blocks of the Dome of the Rock. Its plan is formed by two squares of equal size and with the same centre, superimposed, but with one square turned 45 degrees to the other. This produces an eight-pointed star or octagram, the inner form of which is the octagon, above which rises the drum and dome.

The numbers enshrined in this geometry, and the meaning evoked by them, are fascinating and revealing and appear to relate to ancient traditions, shared by most religions, that certain numbers represent certain spiritual or cosmic qualities. For example, the number four – the square – symbolizes the material world, while the circle or sphere – a perfect geometric shape without beginning or end – expresses the spiritual realm. So at one level, the Dome of the Rock, its hemispherical dome sitting upon a base formed by two squares, is the fusion of the worlds of man and God.

It seems there are additional meanings. The building's fifty-two windows, its twelve internal inner columns and the four piers that divide them, and the seven arches on each of its eight external walls suggest the weeks and months of the year, the four seasons, and the days of the week – so, a calendar.

But this is not the most surprising thing about the interior. The walls are clad with delicate tile-work displaying exquisite geometrical designs, yet they frame a huge and rough-hewn lump of rock that forms the floor. It is this rock that the Temple of Solomon is all about – indeed, what much of the Jewish and Muslim faiths are about. It is regarded as the tip of the intensely sacred Mount Moriah. For the Jews this rock is the foundation stone of God's creation, the junction between heaven and earth and the site of the first Temple's Holy of Holies. It is also believed to be the stone on which Abraham planned, at God's bidding, to sacrifice his son Isaac (or for Muslims, Ishmael). It is the tip of this rock – the would-be sacrificial altar – that is now covered by the Dome of the Rock. In addition, Muslims also believe this is the rock from which the prophet Muhammad ascended to heaven, leaving behind a faint impression of his foot.

Beneath this rocky surface is the building's final surprise: a grotto known as the Well of Souls, where according to Islamic belief, souls will gather at the End of Days to be weighed for their good and evil acts. This building is, I suppose, a countdown to the Resurrection. This is nothing else like it on earth.

49

MALWIYA
Samarra, Iraq | c. 848–52 CE

The Malwiya is, in its way, utterly perfect. It is a vast, tapering, spiral ramp, wrought of mellow red bricks, that rises to a height of 52 metres (170 ft) with utter geometric precision. It stands next to the long brick walls of the huge ninth-century Great Mosque of Samarra (all but its outer walls were destroyed in 1278 when the Mongol Hulagu Khan devastated the city) and towers above the surrounding buildings, seemingly a man-made and geometrical evocation of a holy mountain. Or perhaps it is a structure inspired by the region's ancient ziggurats, a reconstruction of the Biblical Tower of Babel, or a diagrammatic representation of a journey from earth to paradise.

It was built as a minaret when such structures were still largely a novel idea. Prayer in Islam is a private affair – just a mat, the correct orientation towards Mecca, and a righteous and faithful heart are required. Mosques, when they started to appear, were based on no Koranic text. Some were inspired by Old Testament descriptions of Solomon's Temple, a structure that Jews, Christians, and Muslims agree was created by man under divine guidance. But most mosques were inspired by the design of Muhammad's house in Medina, which was a typical courtyard house of the region and period. Early mosques did not have minarets, but at some point in the early eighth century these structures appeared. The earliest surviving minaret, said to have been started around 725 CE but not completed until 836 CE, is attached to the Mosque of Kairouan in Tunisia. It is square in plan and rises through three stages of decreasing size.

Perhaps minarets were inspired by the bell towers attached to Christian churches, and essentially they had the same purpose: to summon the faithful to prayer and to point aspiring fingers to heaven and mark the location of a sacred building, a house of God. In Arabic minaret means 'lighthouse', and they are often referred to as gates – or stairs – between heaven and earth.

Given its early date, the design of the minaret at Samarra can be seen as experimental. When it was designed, the conventions governing minaret design had yet to be established, and when minarets started to appear in large numbers they soon evolved very distinct regional characters. But few minarets are like that at Samarra – although it did have some direct and almost immediate progeny, notably the spiral minaret at the Mosque of Ahmed ibn Tulun in Cairo, built *c.* 876–79.

The geometry of the Malwiya is elemental and basic; the spiral is generated by the use of contiguous right-angled triangles, with the diagonal hypotenuse being the basis for the next triangle in the series. This root-two spiral – known as the spiral of Theodorus after the Greek mathematician of the fifth-century BCE who seems to have first generated it – would have been well known in the Arab world of the ninth century. It is only possible to speculate about its meaning, but any circular or spiral form generated from a square would have been seen to symbolize the immaterial merging from the material, for the square by tradition represented the world, while the circle – with no beginning and no end – represented the divine. And the spiral appears in many sacred buildings, some known in the Arab and Muslim world, and some not – for example, on Sumerian buildings in Uruk, south Iraq, that are perhaps 5,200 years old, and also within Newgrange in County Meath, Ireland, also around 5,200 years old, where the spirals unite to form triskele, or triple-spiral, patterns. The meaning of the spirals on these ancient structures remains unknown.

To ascend the shallow steps of the ramp of the minaret at Samarra is to go on a journey – to the past, to paradise. It's also a reminder of how difficult it would have been to construct, with such perfection, this seemingly simple but continuously curving structure. And the views from it are sensational. First there is the vast enclosure defined by the walls of the adjoining mosque, revealing the audacious size of this now largely lost mega-structure, said to be the largest mosque ever built. And then the layout of the old city gradually becomes clear. It was an ideal world, with a regular street pattern, palaces, and paradise gardens, founded in 836 by the Abbasid caliph al-Mu'tasim as his capital and extended in the mid-ninth century by al-Mutawakkil, who commissioned the great mosque and minaret. At that time, it must have been the greatest – certainly the most orderly and beautiful – city in the world, but in 892 the court moved back to Baghdad and the city fell into gradual decay.

Arrival at the top of the minaret offers a reward beyond a spectacular prospect. There is a kiosk embellished with pointed arches, perhaps added later but if not a poignant reminder that the pointed arch – a key element of medieval European Gothic – had its origin, centuries before the Gothic, in the Islamic world.

50

MAUSOLEUM OF ISMA'IL SAMANI
Bukhara, Uzbekistan | 9th–10th century CE

This mausoleum is, in its perfection of form and symbolism, one of the most satisfying – if enigmatic – structures in Central Asia. With its cubical base supporting a perfect hemispherical dome, it provided a compelling model for succeeding generations of Islamic tombs. But its construction with fine kiln-fired brick and much of its abstract ornament evoke memories of the region's ancient and pre-Islamic cultures.

The mausoleum was constructed between 892 and 943 CE for Isma'il ibn Ahmad, the Samanid amir of Transoxania, who ruled the region and died in 907. During his rule Bukhara was transformed into one of Islam's greatest cities, attracting scholars, artists, craftsmen, and architects. The mausoleum is a miniature monument to the lost architectural wonders of Bukhara's early golden age.

The brickwork is superb, wrought of long, thin, hard, buff-coloured brick slabs that are sometimes used whole and flat but more often cut and arranged to form a wide variety of abstract ornament. The exception is the dome, where the bricks are used – as stretchers only, with no headers – to form a perfect, self-supporting, and gently curving structure. This assured use of fine kiln-fired bricks, the longevity of which reveals a high level of technical mastery in ninth-century Bukhara, must be a legacy of the region's deep-rooted traditions of brick construction. Arguably in Mesopotamia, around 6,000 to 7,000 years ago and about 3,000 kilometres (2,000 miles) to the west, man first started to build with kiln-fired and sun-dried brick. This technique, vital in a region short of building stone and timber, soon took hold and spread, making brick the standard building material of the region.

The meaning of the mausoleum's geometry was no doubt traditional when the structure was designed. The cubical base celebrates the number four, which was seen as an expression of the material world – for example, the four elements and the four seasons, and in Islam the four sacred months and four Holy Books. So it commemorates the material means through which the immaterial was made manifest. The dome, with its circular base, was an ancient symbol of the spiritual.

But the meaning of the abstract decoration is far more mysterious, perhaps of ancient origin when the mausoleum was constructed and now lost in time. The engaged and capital-less columns forming the four external corners of the cube are, along with much of the surface of the cube, textured with a repeating pattern formed, generally, by three horizontal bricks supported by three bricks vertically laid in a gently curving configuration. This gives the elevation a slightly perforated appearance. Other details achieved by bricks – either cut or laid at angles – are rows of small circles and a dog-tooth texture. Some of these decorative and construction techniques might have been inspired by the astonishing and pioneering late eighth-century, brick-built Mosque of the Nine Domes (Masjid-i-Tarikh) at Balkh, near Mazar-i-Sharif, Afghanistan. Certainly the surfaces of the stout columns in the now ruinous mosque (all its domes have collapsed) have a perforated appearance similar to the angle-columns of the mausoleum. But what is astonishing about the exterior of the mausoleum is the row of slightly pointed arches placed at the top of the cube. Each arch is framed by a pair of short columns that vary in their surface decorations, with pairs of columns bearing a spiral decoration alternating with pair of columns featuring a chevron decoration. What can these abstract patterns have meant to the designer of this mausoleum? In sacred buildings, nothing is the result of whim or chance. There must surely have been a connection back to the identical abstract imagery used perhaps 4,000 to 5,000 years earlier – wrought in kiln-fired clay cones – on the walls of sacred buildings in the ancient Sumarian city of Uruk in Mesopotamia. These particular forms, inspired by the geometry of the circle and the square, must in some way have been perceived in ninth-century Bukhara as the elemental language of the sacred, universally applicable in the design of shrines, mosques, and tombs.

The interior of the mausoleum has a richer vocabulary of decoration, including lozenges, discs, and quatrefoils wrought in kiln-fired clay, with columns framing the squinches that reconcile the cubic base of the building with the curve of the dome.

Much that was to distinguish later Christian Romanesque and Gothic architecture and the design of mosques lies latent in this diminutive structure. For example, spiral and chevron patterns adorn some of the most distinguished examples of England's late eleventh- and early twelfth-century Romanesque architecture – notably in Durham and Canterbury cathedrals – and can also be seen on early minarets in the Middle East and Central Asia, and the quatrefoil became a standard detail in Gothic construction. How could these motifs have travelled? One answer is the Silk Road, beside which the Mausoleum of Isma'il Samani stands and along which travelled, in ancient times, ideas as well as goods.

51

CHAPEL OF ST JOHN
**White Tower, Tower of London, London, England |
c.1075–95 | Probably Gundulf, Bishop of Rochester**

The White Tower, started *c.*1075, was built to impress, intimidate, and control. It was a tangible architectural expression of the Norman invaders' determination to hold their conquest.

William of Normandy had gambled much on his military invasion of 1066, and his victory at the Battle of Hastings gained, but did not secure, the English crown. To hold the kingdom – won by a single brilliant act of violence – William turned to architecture. William seized strategic locations and lands from Anglo-Saxon families and granted them to *arriviste* Norman lords, who secured their prizes by means of castle construction. He also promoted Norman control over the hearts and minds of the native population by encouraging French-based religious orders to settle England and, through monastic buildings, disseminate Norman power.

The construction of the White Tower was part of this policy. It was almost certainly designed by a Norman Benedictine monk named Gundulf, whom William soon established as Bishop of Rochester. Anglo-Saxon lords had not built vast stone castles to hold their lands, so the White Tower would no doubt have shocked and awed Anglo-Saxon Londoners – and this was just what it was meant to do. Londoners were particularly hostile to the Normans, and the tower made it clear that a new order had been established.

The site chosen for the tower was critical. Located on the east side of the city, it occupied high land that dominated the passage along the river and the surrounding area. When William constructed the tower, he realized it would be the key to the realm – he who held the tower would hold London, and he who held London held the land. So the White Tower was designed as a cutting-edge example of late eleventh-century military architecture. But it was more than just a fortress and a heart of military dominance; the White Tower was also a palace. As early as the twelfth century it was described as *Arx Palatina*, or fortified palace.[3] So as well as being a potent symbol of William's power, the tower had also to possess a sense of state, luxury, and divine authority. It had to help legitimize William's kingship by suggesting that his military triumph was an expression of divine will, and that the Battle of Hastings had been a trial by ordeal conducted under God's watchful, just, and omnipotent eye.

Central to the White Tower's role as palace and sacred expression of William's right to rule is its most notable and glorious architectural feature, the Chapel of St John's – one of the earliest, most intriguing, and least-altered ecclesiastical interiors in England. Various details inside the tower proclaim its dual role as strong fortress and comfortable palatial quarters, notably the pioneering fireplaces in upper rooms that, with hearths and flues located within the thickness of the walls, allowed individual private chambers to be heated. But the presence within the White Tower of a Chapel Royal – double-height and nobly vaulted – confirms its palace status. And as well as speaking of majesty, the chapel appears to carry a specific message about William's kingship.

The dedication of the chapel to St John the Evangelist is highly significant. In William's age it was believed that St John, one of the twelve Apostles, was the same saint that had written the apocalyptic Book of Revelation, the enigmatic Biblical text that prophesies the end of the world, and the world to come. And the Book of Revelation offers a remarkable insight into the design and meaning of this chapel, suggesting that the Bible had been Gundulf and William's design guide. For example, it contains twelve columns, perhaps representing the 'twelve foundations' of the 'New Jerusalem' as described in the Biblical text.

At the west end of the chapel was a dais on which sat the king and queen. The sacred geometry of the chapel is a permutation of *ad quadratum*, a geometrical system based on interlocking squares generating forms such as the octagon and the oval *vesica piscis*, and within this geometry the royal dais lies exactly opposite the altar. So the chapel is, with its octagonal geometry, like both a baptistery – a place of rebirth – and a finely balanced set of scales. On one side sat the monarch, representing worldly power, and on the other was the altar, representing spiritual power.

Could it be that William conceived the Chapel Royal, in the White Tower, as the heart of his new kingdom, as the New Jerusalem in miniature, the place in which to keep his new kingdom in balance? Certainly the creation of a heroic chapel within his new London palace was, at the very least, a thanks offering to God for the kingdom that William had won in desperate battle and that he believed he could retain only by God's will.

52

CATHEDRAL CHURCH OF CHRIST, BLESSED MARY THE VIRGIN AND ST CUTHBERT OF DURHAM
(Durham Cathedral), Durham, England | 1093 to c.1135

Durham Cathedral, intensely atmospheric and architecturally pioneering, is a key transitional building. It is the epitome of robust Romanesque architecture, which was the international style of Europe in the eleventh and early twelfth centuries, inspired by the remains and culture of Rome; but it is also, in some of its structural innovations, a harbinger of the coming Gothic. It is a mighty bastion of the Christian faith, and in many key aspects its design was seemingly inspired by Biblical texts, yet it also reflects, in its forms and details, the influence of Islam. This faith – in a sense a logical reformation of Judaic and Christian liturgy – became familiar in the West due largely to the First Crusade of 1096, resulting in the capture in 1099 of Jerusalem, with its masterpieces of Islamic architecture such as the Dome of the Rock. The reconstruction of Durham Cathedral started in 1093 and had reached a critical stage by 1100, when ideas from the East started to arrive and – arguably – have a significant influence on design and construction.

In 1071, William I made the bishopric of Durham like no other in the land. The strategic importance of Durham in the Normans' continuing efforts to secure their conquest led to the bishop being given the powers of a head of state. He could raise his own troops and mint his own money, all to give him the power to hold and control the region. It was within this context that a Norman cleric and Benedictine monk, William de St Calais, became prince-bishop in 1081 – and continued the construction of monastic buildings in Durham that culminated with the rebuilding of the cathedral, in spectacular manner, from 1093.

As was usual, construction started at the east end so that the choir and high altar could be completed as quickly as possible and the cathedral used for worship. But before this, the design – or at least the basic plan – for the entire cathedral must have been resolved. As was typical for the time, a basilica plan (a central passage separated by columns from flanking aisles) was used for the nave and choir. The geometry underpinning this design is the subject of much speculation but seems to have been the usual *ad quadratum*, based on squares that overlap or turn 45 degrees to each other to create octagons and eight-pointed stars. All of this was highly charged with sacred and symbolic meaning, relating to rebirth and the harmony between celestial and terrestrial life. The geometry seems to have defined and unified the cathedral and monastery as a heavenly world, inspired perhaps by the Book of Revelation that describes the 'New Jerusalem' as a mighty cube.

The designation of the cathedral and its monastic buildings as sacred ground was particularly important since they were home to the shrine of the Anglo-Saxon cleric St Cuthbert, who was regarded with much devotion as the patron saint of Northern England. So Durham was a centre of popular pilgrimage; this was profitable but could also, the French clerics realized, be used to unite the Norman newcomers with the Anglo-Saxon community if the local saint was treated with respect and re-housed in style.

The choir is a typical piece of Romanesque architecture, with walls – thick and pierced with relatively small round-arched windows – designed to carry most of the weight, and restrain the outward thrust, of the stone vault. This vault was started in around 1100. It is a groin vault, which means it's formed of two semi-circular vaults intersecting at right angles to each other. The divisions between the bays of the vault are marked by semi-circular stone ribs, but – new for England at the time – the vaults are also stiffened by diagonally placed transverse ribs. These gathered much of the outward thrust of the vault and took it to areas of the walls reinforced by stout, buttress-like pilasters.

The choir vault was replaced a hundred years after construction, leaving the north transept vault, completed in about 1110, as the earliest high-level vault in Britain. This vault is fascinating because within it the structural principles of Gothic architecture – with its integrated system of ribs, piers, and buttresses – are starting to emerge, and this is well before the conventional 'birth' of Gothic construction in 1144 with the choir of St Denis in Paris.

The pioneering Gothic spirit of Durham Cathedral took more tangible expression in 1128 when the nave vault was started. It incorporates pointed arches – the emblem of the Gothic – that permit a more finely engineered and poised structure. These pointed arches, spanning the width of the nave and used in conjunction with rudimentary flying buttresses (concealed in the triforium), relieved the walls of part of their structural role, making possible the insertion of large windows.

The famed columns of Durham – in choir, transept, and nave – hint at another remarkable story. In their stout girth and vast load-bearing capacity they are far from elegant and are, from the structural point of view, quintessentially Romanesque. But the abstract patterns they bear seem ancient and remote in origin. The columns and the patterns could be inspired by Biblical descriptions of Solomon's Temple, making the cathedral yet one more medieval evocation of this seminal building that was believed to be divine in origin. The columns in the choir, constructed around 1095, are incised with spirals, as are all but one of the transept columns, built from c.1099 to 1104. The exception combines spiral with chevron and appears to be the centre-point of the sacred geometry of the cathedral and its precinct. The columns in the nave, started in the early 1120s, have circumferences equal to their height (so are essentially squares rolled into cylinders), and the patterns with which they are incised include spirals, chevrons, and lozenges. The meaning of these patterns is much debated, but they seem to be a timeless prayer in stone – the language of the sacred – brought back from the Muslim architecture of the Holy Land, and an echo of the 4,500-year-old sacred architecture of Mesopotamia.

53

CHARTRES CATHEDRAL
France | 12th and early 13th centuries

The cathedral is, in many ways, the epitome of Gothic sacred architecture. It is a structure that makes manifest man's deepest beliefs in the immaterial world, the world of the spirit. Chartres, with its powerful proportional systems and its very distinct sculptural decoration, remains a deeply mysterious structure.

A church was constructed here, over an ancient sacred grotto and well, soon after Christianity came to the region. This was replaced from 1020 by a mighty cathedral that was engulfed by fire in June 1194, with only the west front, which had been completed in about 1160, and the ancient crypt surviving. Rebuilding started immediately and by the late 1220s was virtually complete.

In twenty-five brief years, the art of Gothic design and construction – still tentative in 1194 – reached speedy and marvellous maturity. In the nave and aisles of Chartres, the structural potential of the Gothic pointed arch is utilized to gain high vaults and wide bays, and the Gothic structural system of ribs, piers, and flying buttresses is fully developed.

This system is used to create an elegant structure – a house fit for God – in which the outward thrust of the high vaults and arches is carried to the ground and neutralized in the most elegant manner imaginable by being met by the opposite and equal counter-thrust exerted by the flying buttresses. This ingenious skeletal structure, with loads carried on arches, stone ribs, and buttresses, leaves the interior light and open and relieves the outer walls of much of their structural role. This allowed them to be pierced by large windows through which God's light – manipulated by emblematic stained glass – could illuminate the interior of the church and the minds of the worshippers gathered within.

The identities of the cathedral's designers and engineers, and the source of the new and complex structural ideas used in its construction (and, indeed, of the money required to build such a high-quality structure at such speed), remain shrouded in mystery.

When virtually complete in the 1220s, Chartres became one of the wonders of the world. Pilgrims flocked there to see its great relic, the Veil of the Virgin, but also surely to see the architecture, unprecedented in its lofty and minimal elegance. They would have seen it as an homage to the power of God's creation, as an earthbound image of the celestial city, a building that was the Bible built in stone and a reconstruction of God's house in Jerusalem, the Temple of Solomon.

All of this is implied – or confirmed – in the design, proportions, form, and details of the cathedral. The most westerly door of the porch, incorporating an image of King Solomon in judgment, is a clue that the building's proportions are inspired by the Biblical descriptions of the Temple of Solomon.

The west front of the cathedral includes the royal portal, completed in about 1155, and which survived the fire. It contains three doors. The most northerly is the Gate of the Ages, showing the Ascension of Christ and signs of the zodiac, alternating with images showing the works of man. The south door is the Gate of Birth, presided over by the Virgin Mary. This includes the image of two men, twins, sheltering behind a single shield – perhaps another zodiac sign, Gemini, or an image representing the religious military order of the Knights Templar, who, based on the Temple Mount in Jerusalem, were almost certainly involved in this inspired recreation of Solomon's Temple in the flat fields of northern France. The main, central door shows Christ sitting in judgment within a pointed oval called a *vesica piscis*. This is a perfect geometrical shape formed by the overlapping of two circles of equal size. In the Middle Ages this geometrical form was perceived to possess deep power and meaning, and it was employed in the design of churches and in painting and sculpture. It was a representation of the Virgin's vulva, the sacred passage through which the Saviour entered the world.

Inside the cathedral, light is all-important. Chartres is virtually designed around its windows, with their richly coloured stained glass; this is the best collection of early stained glass in the world, and it bathes the interior in a beautiful, mystical, and fiery light. On all four cardinal points of the compass are vast, circular rose windows containing geometric patterns imbued with once powerful, though now uncertain, symbolic meaning. The rose is one of the symbols of the Virgin (to whom the cathedral is dedicated), so these could be dedications to her, and each is organized around the number twelve, and so presumably is to do with time – the hours of the day or the months of the year.

The interior possesses astonishing power. As the Bible makes clear, it is not the seen but the unseen world that is significant, and in this cathedral it is not the walls that are of prime importance – no matter how glorious their stones might be – but the space they define. God lives in space, in light and colour: 'look not at the things which are seen, but at the things which are not seen: for the things which are seen are temporal, but the things that are not seen are eternal' (2 Corinthians 4: 18).

Like all great sacred buildings, Chartres deals with the big ideas that have obsessed mankind through the ages – life, death, morality, where we come from and where we are going to. There is, in a sense, a universal religion and a quest for eternal truths, and Chartres Cathedral is in many ways a spell-binding and uplifting expression of this universal religion and universal culture.

54

MINARET OF JAM
Ghor Province, Afghanistan | c.1174 | Ali ibn Ibrahim al-Nishapuri

Afghanistan was at the crossroads of ancient civilizations and 2,200 years ago at the centre of the known world. The land became rich, and an incredibly diverse melting pot of peoples and cultures. This has given Afghanistan a vast and complex heritage, for it was not only goods that passed along its trade highways – such as the Silk Road – but also ideas, religions, and art.

I went to Afghanistan in 2007 on a very specific mission: to view one of its most remote and architecturally important treasures, the almost mythic twelfth-century Minaret of Jam. Though few have seen it – indeed, it was only rediscovered by the outside world in the 1950s – it is one of the finest Islamic structures in the world.

It is also one of the most threatened, a victim of its remote and wild setting and the decades of conflict that have benighted Afghanistan. The minaret, one of the tallest in the world at 60 metres (200 ft), has suffered centuries of neglect, has had its foundations undermined by the river on which it stands so that it now leans at a precarious angle, and in recent years has been a victim of looting. UNESCO, which in 2002 declared the minaret and its setting a World Heritage Site, has carried out some emergency repair work and constructed a protective river wall.

I travelled from Herat as part of a convoy of fifteen vehicles packed with about seventy heavily armed Afghan policemen, the escort for the BBC team of which I was part. The journey was long and stressful, but after nearly eighteen hours – including a few hours' rest in a remote village – I caught my first glimpse of the minaret. It was no disappointment.

The minaret rose tall in a narrow valley flanked by steep mountains – an image of delicate man-made perfection set amongst, and in dramatic contrast with, rugged and sublime nature. It's a dramatic memorial to the long-since-obliterated Ghorid Empire, which rose to power in the mid-twelfth century, burnt bright for only a few decades, and within seventy-five years all was over. At the height of its glory the empire embraced modern Afghanistan and Pakistan and

stretched as far south as Delhi in India, but in the early thirteenth century the empire was weakened by internal strife
and finally collapsed when attacked in the 1220s by the
Mongol warriors of Ogedei and Genghis Khan.

The Minaret of Jam not only reveals the exquisite architectural taste and advanced engineering skills of the Ghorids
but also, some argue, marks the site of their long-lost and
once-famed summer capital of Firozkoh.

But beyond all of this, the minaret has another meaning
– it carries a message, written in its very fabric. It's constructed
of well-wrought, regular, and very hard kiln-fired bricks of a
mellow yellow colour. They are beautiful and very sound,
and in themselves no mean technical achievement in this
remote location over 800 years ago. And it is in these bricks
that the message of the minaret is inscribed. Its lower portion
is covered with ornate geometric patterns incorporating
eight-pointed stars and other Islamic emblems, and woven
into this all-enveloping pattern is Kufic text – in fact, the
entire nineteenth Sura of the Koran. Entitled Maryam, this is
a very particular Sura, for it tells of the Virgin Mary and Jesus
Christ, figures greatly esteemed in Islam, and of those proph-
ets – such as Abraham, Isaac, Joseph, and Ishmael – that are
venerated by the three religions of 'The Book': Judaism,
Christianity, and Islam. Essentially the Sura is a reminder of
what the three religions have in common, a reminder that
they have a shared root, similar aspirations and ethics, and
similar vision of a day of judgment when all will ultimately
be held responsible for their actions on earth, with paradise a
promise for the righteous. So it seems to me that the placing
of this text on the minaret represents a quest for tolerance
between the people of the three religions; it's an appeal for

harmony and understanding that must surely reflect the aims of the Ghorid rulers and which, in these divided times, is more relevant than ever.

The minaret also contains other texts: at low level, in a damaged foundation panel, is the date of construction, now thought to suggest between late 1174 and late 1175. The name of an architect is also mentioned – Ali ibn Ibrahim al-Nishapuri, whose name suggests he was Persian in origin. In the centre of the minaret, in large blue-glazed lettering, is the name of the ruler during whose reign it was constructed, Ghiyath al-Din.

Observing the long shadow of the minaret in morning light, I suddenly saw it as a mighty gnomon marking this as a sacred sight, a centre of the known world when the Ghorids ruled the region. It is undoubtedly one of the most beautiful and miraculous creations of the Islamic world, and also now one that we must fight to save.

55

ANGKOR THOM
Siem Reap, Cambodia | Late 12th century

The Khmer Empire that created the monuments of Angkor flourished from 800 to 1400 CE. It commanded the productive lowlands of Cambodia and achieved a surplus in rice – and with plenty, came leisure, the arts, and the time to evolve a complex setting and symbolism for the Hindu religion that had arrived from India. The golden age of the empire was in the twelfth century, and during this period a series of stunning structures was created over a large area, currently calculated at 1,000 square kilometres (400 sq miles). The best preserved and best known is Angkor Wat, whose name means 'temple city'.

All of this was intended to form a sacred landscape for the rulers and people of the empire – a vision of heaven on earth. The intention and meaning of the place was clear to all who visited. Angkor Wat was dedicated primarily to the god Vishnu, conceived, like all great Hindu temples and Buddhist stupas, as a model of the universe with the sacred Mount Meru at its centre, as the dwelling place of the deity. Construction started in about 1140, and the building was intended to serve both as temple and tomb for its creator, the *devaraja* or god-king Suryavarman II, who died in 1150 just as Angkor Wat was completed.

Adjoining Angkor Wat is Angkor Thom, meaning 'great city'. This is a massive walled city, rectangular in form and 9 square kilometres (3.5 sq miles) in area, that was started in 1181 by King Jayavarman VII. It is a remarkable conception that was the product of turbulent times. The Chams, a warlike people centred in Vietnam, captured Angkor in 1177, but were quickly defeated and expelled from Cambodia. Jayavarman built the walled and fortified city of Angkor Thom both as a symbol of his triumph and an attempt to ensure that this 'great city' could never be taken by an enemy.

The walls of the city are pierced by five main gates, each of which – as a foretaste of things to come – has an upper portion incorporating towers that take the form of human faces. These stone-wrought faces, with their enigmatic smiles, are the world's best-known examples of anthropomorphic architecture. They can be perceived as mystical and, like the faintly smiling Great Sphinx, suggest that in some way, there is a riddle to solve. But more prosaically, they proclaim a change that took place in the empire in the late twelfth century. Hinduism was replaced as the state religion by its child, the more philosophical Buddhism; this had emerged from the teachings, around 2,500 years ago, of the Hindu prince Siddhartha, who is said to have achieved enlightenment – to have grasped the truth of the human condition, of the past, and of the future – through meditation. The faces at Angkor Thom bear the Buddhist smile of calm but ecstatic enlightenment, of perception.

At the centre of Angkor Thom survives its most important and most impressive building, the Bayon. Also started in 1181, this spectacular tomb and temple surrounded by a moat is a monument not just to a man but also to Angkor Thom's new religion.

Initially at least, the faith of Buddhism enjoyed an easy relationship with Hinduism. As Buddhism had evolved out of Hindu theology, so it seems Buddhism in Angkor had evolved – and merged with – Hindu beliefs. Everywhere in Angkor Thom, images of the two beliefs overlap. But tensions did grow, and in the late thirteenth century, in a return to Hinduism, Buddhist images at the Bayon began to be replaced or destroyed.

Jayavarman saw himself as the compassionate Mahayana Buddha, so the faces on the city gates and the tiers of serenely smiling heads that surmount the Bayon are naturally images of the compassionate Buddha, or rather of Jayavarman in the guise of the compassionate Buddha, implying that he was a bodhisattva, an enlightened being committed to helping others.

But it seems that, as well as being a tomb and temple, the Bayon has another meaning. It is said to have incorporated fifty-four carved heads, and since there were fifty-four provinces in the Angkor Empire, the Bayon can be read as a model of the empire itself. So as well as being sacred, the architecture of the Bayon also carried a political message and was an image of both divine and worldly power – especially when the long-lost crowning gigantic image of Jayavarman as a bodhisattva towered over the other heads to form the central pinnacle of the structure.

To modern eyes these huge faces, with their haunting smiles, can take on additional meaning. When viewed in failing forest light, the faces, whose expressions become mobile when reflected in the rippling surface of the moat, seem alive, suggesting that the temple and the king it celebrates are eternal living beings. Perhaps this always was the intention.

56

ROCK-CUT CHURCHES
Lalibela, Ethiopia | Early 13th century

King Lalibela, after whom the town of Lalibela was named, was the inspirational monarch who reigned from the late twelfth century and who conceived the powerful political and religious idea of declaring his realm the new Holy Land, with his capital the New Jerusalem. Christian missionaries had almost certainly entered what is now Ethiopia in the early fourth century, and they brought the Orthodox brand of the faith, which remains the religion of the land. So by the time

Lalibela came to the throne, Orthodox Christianity was the deep-rooted traditional religion of the region.

Lalibela cannot have changed much in the last 800 years or so. If anything, it has got smaller and more primitive – certainly far more impoverished. The town is divided, as in Lalibela's time, by a small and meandering watercourse. Now it is mostly dry and clogged by garbage, but 800 years ago it probably flowed more freely – certainly enough to inspire Lalibela to name it the 'River Jordan' as part of his scheme to make his city, the new 'Holy Land', a valuable place of pilgrimage. And key to this route of pilgrimage are the rock-cut churches, excavated on each side of the 'River Jordan', that represent sacred sites and places in the authentic Jerusalem and Holy Land.

King Lalibela's vision is extraordinary, and still far from fully understood. It is not really even clear what the ages of the churches are, or even if they all started life as churches. Some, according to recent analyses, could be as old as the sixth to eighth centuries and so were pre-existing structures that were adapted and incorporated to play a role in Lalibela's grand design. Some may have started life as houses or refuges.

And then there are the deep trenches, cuttings, and long and winding rock-hewn tunnels that surround and connect some of the churches and sacred sites. Some are, of course, the consequence of construction, the result of excavating the church from the rock, but many others are mysterious. Perhaps they are the remains of ancient defensive systems, or of pilgrimage routes that allowed pilgrims to move from church to church in a state of uninterrupted religious bliss.

As well as creating revenue and prestige through pilgrimage, it is possible that the creation of this New Jerusalem was an attempt on Lalibela's part to legitimize and consolidate his rule. The earlier Ethiopian kings of Axum claimed their right to rule, according to the ancient theological text of the Kebre Negast, through their descent from Solomon's son Menelik I, and that gave them, through Solomon, a lineal connection to the House of David and to Christ. Lalibela lacked this pedigree, so he perhaps hit on the cunning plan of bringing the Holy Land to Ethiopia, through his massive and sacred work at Lalibela, in order literally to build his Zagew dynasty into the fabric and theology of the land.

The duplication of the Holy Land was meticulously planned and executed. On the north bank of Lalibela's 'River Jordan' is the recreation of the terrestrial Holy Land. Here is the Beta Medhane Alem ('House of the Redeemer of the World'), the Beta Maryam (House of Mary), and the Beta Golgotha and Beta Mikael, commemorating the crucifixion. On the south are churches that represent the celestial Holy Land, including Beta Emmanuel, which represents the Seven Heavens, and Beta Merkorios and Beta Gabriel-Raphael, which, within their shared cavern, possess a nearly inaccessible cave that is said to represent paradise.

In addition to these churches there are sites and caves marking other holy sites that include the Mount of Olives (here called Debre Zeit), Bethany, Mount Tabor, and Bethlehem. In addition to the ten Holy Land churches made or remodelled by Lalibela, there is one more, created by his widow as a memorial to him after his death in about 1220. This is St George, in many ways the finest, the most emblematic and sculpturally perfect, of the churches.

St George stands in a 20-metre (65-foot) deep cutting from which it was excavated. It is a Greek cross in plan, with the upper face of each arm of the cross more or less square in plan and equal in size to the square that forms the roof over the centre of the church and from which the arms of the cross radiate. The church is roughly twice as high as it is wide; it's like a pile of cubes, one upon each other. This church, apparently the last created, is the culmination of Lalibela's heady enterprise to create the new Holy Land and to establish his family as the new Solomonic dynasty. Perhaps executed by his widow to help secure the family in power, it is, in its abstract way, the new Temple of Solomon for his New Jerusalem. It is a Greek cross in plan but a mighty cube in spirit, echoing Biblical descriptions of sacred architecture. This church, like the New Jerusalem itself, is the world of spirit made manifest in cubical geometry. What ambition.

57

SRI RANGANATHASWAMY TEMPLE
Srirangam, Tamil Nadu, South India | c.13th century

The temple town of Sri Ranganathaswamy, located on the sacred island of Srirangam in South India, dates from the thirteenth century and is dedicated to an incarnation of Vishnu, one of the three most powerful Hindu gods. The temple town, 23 hectares (56 acres) in extent, is one of the largest and architecturally finest evocations of a Hindu vision of paradise and lies next to the Cauvery River, held by South Indians to be as holy as the Ganges in the north.

The Sri Ranganathaswamy Temple is a truly extraordinary place. It's conceived as a giant *mandala*, a geometric and symbolic diagram of the universe, and is formed by seven concentric enclosures that represent the various stages of a journey of the soul. At Ranganathaswamy this journey starts at the worldly outer limits of the temple, with the route leading gradually to the sacred centre – to the source of all spiritual power and knowledge. The object of this journey is to gain unity with god, to achieve *moksha*, the release from the painful and endless cycle of life, death, and rebirth on earth.

The temple is the world in miniature. Within its seven enclosures are markets, shops, workshops, and houses, as well as shrines and prayer halls – but the uses within each enclosure become increasingly more sacred as the epicentre of the temple is approached.

Each enclosure is of rectangular shape with, roughly in the centre of each of its four sides, a pyramidal entrance tower called a *gopura*. The alignment of the *gopuram* – marking the routes through the temple that all converge at the central sanctuary, or *vimana* – has a most dramatic visual effect. The *gopuram* themselves are architecturally striking, each loaded with tiers of brightly coloured sculpture showing gods and goddess, worldly rulers, ferocious demons, and protective spirits.

As with all sacred buildings, nothing here is without meaning – for example, why seven enclosures? In Hinduism, the Puranas, the sacred texts that describe the beginning and end of the world, say the cosmos consists of fourteen realms, comprising seven 'visible' upper worlds and seven nether-worlds. Also, a significant clue to one of the meanings of this temple, Hindus believe there are seven elements or dhatas of the human body, with the *atman*, or soul, dwelling at its centre, and that the body has seven layers of skin and seven 'centres'. And this temple town is not just a vast place of worship created by man, it is also for Hindus a representation of the body of the god for which it was created – a divine being, animated through worship.

The second enclosure contains tranquil streets lined with houses of the Brahmin families that, by tradition, serve the temple, but it is at the fourth enclosure that the temple proper starts. Profane buildings and uses decrease, while the sacred structures related directly to the function of the temple increase in number, scale, and quality. These include the Temple of Venugopala – one of the best architectural essays in the temple – dedicated to Krishna, the most powerful incarnation of Vishnu. Krishna, like Christ, is said to be a divine saviour who preached mercy and love. Unlike Christ, however, Krishna expressed his love in many forms and is said to have had 16,000 wives in addition to his eight principal queens. This is, perhaps, why the exterior of building – said to date from the fourteenth century but could be as late as the sixteenth – is embellished with exquisite carvings of curvaceous and beautiful young women. One plays upon a vina (a kind of stringed instrument), while another stares narcissistically into a looking glass, and yet another lurks naked in a corner trying, not very successfully, to conceal her private parts beneath her hands.

As an eager Brahmin once explained to me, in the temple at Madurai, 'sex life is divine life'. Temples are not only an image of the body of god but a fertility symbol, and in the case of Madurai its plan and decoration express the Kundalini – the coiled serpent, the 'fire-snake' of sexual energy said to reside in the base of the spine – which pilgrimage through the temple is intended to stimulate and transform into explosive spiritual energy. Sex and the divine are, of course, closely related – procreation is mankind's great god-like act of creation – but in Hindu temples sex is a means to spiritual enlightenment, never an end in itself.

The fourth enclosure also contains the 'Thousand Column Hall', in the past the home of the temple dancing girls who would perform to please both gods and men, and whose rhythmic dancing, chanting, and music would create harmony and energy that would help put the earth to rights.

The final two enclosures and the inner sanctuary are open only to Hindus, so all those of other faiths can do is stand and stare into this forbidden world, pondering upon the mysterious 'doorway to heaven' that is located within and the central golden shrine. This, the home of the deity, is the heart and soul of the temple and makes it, for Hindus, a living thing.

58

JANGGYEONG PANJEON
Haeinsa Temple, Mount Gaya, South Korea | 13th–15th century

The Haeinsa Temple was constructed in a high and remote location, with its buildings organized on a series of terraces cut into the gently sloping mountainside. Although now intensely picturesque, solemn, and sacred in appearance, the temple was designed, built, and maintained with one primary functional purpose in mind – a purpose pursued with such brilliant and focused intent that the entire complex is as much a large and well-planned machine as a piece of architecture. The purpose was to store, safe from fire, the elements, and insect attack, 81,258 wooden blocks carved with Buddhist texts. Known as the *Tripitaka Koreana*, the blocks were made between 1237 and 1248, following the destruction of an earlier edition in 1232 by Mongol invaders. They are now revered not just for their antiquity and for the sacred texts they contain but also as outstanding works of art, masterpieces of carved calligraphy, with Chinese characters of regular and beautiful formation. The blocks represent the world's oldest surviving corpus of Buddhist doctrinal texts in Chinese, and with no known faults. The blocks also, with their purpose-built depositories, form what is arguably the oldest intact and fully catalogued library in the world.

The Janggyeong Panjeon depositories at Haeinsa consist of four pavilions – two long and two short – arranged around a narrow, rectangular courtyard. Being of prime importance to the temple, these four buildings are placed at the highest level in the compound, and they are orientated to mitigate the effect of strong, damp prevailing winds. The existing pavilions were constructed in 1398 in the simple style of the early Joseon period. Before this time the blocks were stored until 1318 in the Taejanggyong P'andang, outside the western gate of the Ganghwa Fortress, and then in the Sonwonsa Temple on Ganghwa Island, before being moved to purpose-built accommodation at Haeinsa Temple.

The two long buildings at Haeinsa are of significantly different plan form. One is divided into two large chambers by an entrance passage in the centre of the long sides. The other contains a prayer hall and a corridor running its full length, set against an outside wall. This corridor makes it possible to walk internally between, and enter, the series of chambers in the pavilion. Within these chambers the blocks are stacked, in order, in timber shelves, each with its own number for easy and speedy identification. Being a depository of printing blocks rather than of books, the library does not have – or require – a conventional reading room. But this is no dead repository of history, for the blocks are in constant use, with prints being made from them in traditional manner and dispatched as required.

Each block weighs around 3.25 kilograms (7 lb) and measures 70 × 24 × 3 centimetres (27.5 × 9.5 × 1.2 in). The blocks are made from birch, and the manufacturing process was meticulous and well considered. Each block was boiled in salt water and left to dry slowly for three years, with only the hardest then selected for carving. These chosen blocks, to further enhance their strength, were reinforced with metal. Each block was carved with text on both sides then coated with thick, grey, poisonous lacquer to help preserve it from insect attack.

In addition to this direct protection from insects, the preservation of the blocks during the last 600 years at Haeinsa has been achieved by cunning and thoughtful architectural design that controls the environment within the pavilions and modulates temperature and humidity. These controls appear to have been fine-tuned over the centuries, notably during repairs that took place in 1457 and during the seventeenth century.

The timber posts forming part of the main structure of the pavilions sit on stone pads to protect them from damp, while the floors are made from layers of clay, charcoal, sand, salt, and limestone that reduce humidity by absorbing moisture. In addition, the buildings are surrounded by shallow troughs that rapidly carry away rainwater when it falls from roofs, thus protecting the walls from damp. Added protection comes from carefully calculated natural ventilation, provided by two tiers of windows fitted with timber louvres that ensure airflow while protecting the interior from driving rain. The shelves on which the blocks are stacked are also designed to allow air to flow freely, with warm air rising naturally through the shelves in summer to combat any damp that might have built up during the winter. This well-calculated design has ensured that these pavilions, unheated in any way though located in an areaof climatic extremes, have through the centuries preserved the treasures they contain intact, free from attack by damp, insects, or rodents. The brilliance of the design was confirmed in 1971 when the Korean authorities, in an attempt to better safeguard the blocks, started to move them to modern, heated, and climate-controlled buildings of concrete construction. The blocks that were re-housed quickly started to deteriorate, so they were rushed back to Haeinsa.

59

BASILICA OF SAN LORENZO
Florence, Italy | 1419–59 | Filippo Brunelleschi

San Lorenzo is an ancient church that was rebuilt from 1419 by the pioneering Renaissance architect Filippo Brunelleschi. The patrons were the Medici, whose palazzo stands nearby. The entrance front of the basilica is raw, just bands of very rough and exposed brickwork, because the entrance elevation planned by Brunelleschi was abandoned. The cliff-like and rugged nature of this elevation makes what lies beyond even more surprising. The moment the door is passed, there can be no doubt that this is one of the great architectural sights – and delights – of the world. The vista is one of serenity and symmetry, with all elements ordered with a bold classical clarity and a clear sense of proportion. The main architectural components are emphasized by being executed in beautiful blue-grey *pietra serena*, and this helps to create a powerful perspective – which is hardly surprising, since Brunelleschi had been instrumental in the development of linear perspective drawing and was no doubt determined to use the building to demonstrate the visual power of this new artistic discipline.

Like other artists of the early phase of the Renaissance, Brunelleschi wanted to combine new knowledge with the rediscovery and recreation of the classical past. He wanted to discover and apply the engineering and artistic secrets of the past and make the great age live again to serve the modern world.

Brunelleschi was born in Florence in 1377, and appears to have been trained as a goldsmith, but by 1419 was at the cutting edge of pioneering Renaissance architecture. In that year he started work not only on San Lorenzo but also on the dome of Florence Cathedral and on the Ospedale degli Innocenti in Florence, which, with its uniform arcaded elevation, created a piazza in the manner of a Roman forum and offered a model for the design of uniform urban terraces and squares that remained an inspiration for generations.

With San Lorenzo, Brunelleschi created an idealized version of a Roman Basilica adapted to Christian worship. Tall and correctly proportioned Corinthian colonnades define the central nave, which is flanked by chapels and wide aisles with ceilings lower than that of the nave. The proportional relationship between the nave and aisles was evidently keenly considered from both the visual and hierarchical points of view, with the aisles' bays being square while those of the nave are of 2:1 proportion.

The nave itself is covered with a magnificent and archae-ologically accurate flat, coffered ceiling. Towards the altar end are wide and tall transepts that, where they meet the nave and aisles, define a crossing space. Off the south transept is the Old Sacristy, domed both to symbolize the heavens and, no doubt, as a modest act of homage to the Pantheon in Rome, which Brunelleschi had studied.

The interior of the church is an intensely moving and convincing evocation of antique classical architecture conceived when the Gothic was to flourish in Europe for yet another century and designed a few years before work started on one of Venice's Gothic jewels, the Ca' d'Oro on the Grand Canal. The church is, artistically, far ahead of its time. The light flooding in from high-level windows, the semi-circular arches springing from the tops of the columns, and the clarity of the architectural detail, with all elements superbly inte-grated, give the interior an elegance and a profound simplic-ity. The visual power of the interior would have been even more intense if all had gone according to plan, but lack of funding slowed down the project and Brunelleschi's initial design was diluted and altered, and it was only completed in 1459, over ten years after his death.

60

SANTA MARIA DEI MIRACOLI
Venice, Italy | 1481–89 | Pietro Lombardo

This little church is a perpetual reminder of why it remains a delight to wander through the less traversed backwaters of Venice. Every time I see the church it comes as an utter surprise – even when I'm looking for it – and its cubical, hemispherical, jewel-box perfection always makes it look rather alien and shocking, as if the building were as perplexed to be lurking down a network of anonymous alleyways and minor canals as I am to find it in such company. The physical nature of the building can be described briefly. The secret of its success and the charm of its presence require a little more exploration and explanation.

It is essentially a double cube in form, topped by a barrel-vaulted roof that is simply a cylinder cut long-wise to present a gable pediment of semi-circular form. And at the east end of this double cube is another, but with its long axis vertical to form a tower topped with semi-circular pediments and a small dome rising from a drum. The tower, with its crowning dome, rises above the altar. Most pleasingly, the decorative semi-circular pediments on the tower echo the larger semi-circular pediments on the main body of the church.

This combination of forms – double cubes, semi-circles, and a roof of half-cylinder shape – might seem simple, but in the late fifteenth-century Renaissance such primary shapes were potentially full of meaning. The most direct clue is offered by the mid-fifteenth-century writings of architect and theorist Leon Battista Alberti, who, in his *De Re Aedificatoria*, discussed form and proportion in detail. He explained that Renaissance theory of proportion was derived from the study of exemplary ancient Roman structures and from the observation of nature, which was a sure guide to beauty because it is the handiwork of God. When describing Roman temples – an excellent model for a modern church – Alberti wrote that 'they differ in that some are round, some quadrangular', but all are inspired by nature. Alberti pointed out that 'Nature delights primarily in the circle ... the earth ... the animals, their nests', but also that in many of 'their quadrangular temples our ancestors would [make] a length twice their width'.⁴ So Alberti lists the proportions that Pietro Lombardo used for Santa Maria dei Miracoli, along with their ancient pedigree and divine inspiration.

Evidently the church is intended to be an evocation of a Roman temple and so its details are classical, but Lombardo's classicism is of a most vigorous type, and still imbued with the some of the Gothic spirit of invention and individual expression. The exterior elevations are embellished with two tiers of arcades, incorporating composite pilasters below Doric on the body and Ionic on the tower, again derived from Alberti or from observation of Roman remains such as the Colosseum in Rome.

But as striking as its bold classical design is the material with which the exterior is clad. It is veneered with slabs of marble, their grains matched for stunning visual effect, embedded with jewel-like roundels in varied marbles and with architectural details such as pilasters wrought or clad in lighter-coloured marble. Designed in the round and loaded with carved detail, the church feels almost more like a sculpture than a building. And this is no surprise, since its designer Pietro Lombardo was, like many Renaissance architects, also a sculptor. Lombardo – or a follower – achieved a similar jewel-box-like, rich, and encrusted architecture with the elevation added in the mid-1480s to the Palazzo Dario on the Grand Canal.

The interior of Santa Maria is richer still, the mighty barrel vault coffered, gilded, and restrained by wrought-iron tie bars, the walls marble-clad and with a rich array of carving.

John Ruskin categorized and celebrated Venetian Gothic in his *The Stones of Venice*, a three-volume work on the art and architecture of Venice that was published between 1851 and 1853. Ruskin was naturally critical of the Renaissance classicism that eclipsed his beloved Gothic. But even so he felt compelled to commend aspects of this beautiful little classical church. Naturally, as a great, if doctrinaire, champion of Gothic architecture, Ruskin was critical of the building and condemned its decoration as 'unmeaning'. But he had to admit that 'its … refined … sculptures should be examined with great care, as the best possible examples' of what he dismissed as 'a bad style'. Faint praise indeed, but for any classical building to get any praise at all from Ruskin was a remarkable achievement. Today, of course, when confronted by this glorious little building, it is Ruskin's words that are 'unmeaning'.

61

HANGING TEMPLE
Shanxi Province, China | 16th century

Mount Hengshan is one of China's holy mountains and as a place of tranquil beauty has long attracted temple builders.

Most of these temples and related monasteries were closed, vandalized, or even destroyed during the Cultural Revolution of the 1960s. But during the past quarter of a century, as China has re-embraced some of its diverse religious and regional cultures, many temples have been revived, and the monks have returned.

One of the most architecturally striking temples to survive is Xuankong Si, the 'temple in the air' or Hanging Temple, so named because, as if in defiance of the law of gravity, it hangs off, or rather projects from, a high cliff-face. It started life as a Buddhist temple but gradually incorporated Taoist shrines – a sign of the harmony that traditionally exists between these two ancient religions – and is now, once again, a revered site in the Chinese religious landscape.

Tao is an ancient religion that, in its perceptive and sensitive understanding of nature, has much to teach the modern world. Tao, or Dao, means 'the way', and the core of its belief is that mankind, a product and force of nature, must live in harmony with, and not against, the way of nature – must take nature as a model and learn from it. Taoists strive to follow the principle of wu-wei – of 'not forcing' – and observe that nature works by itself, just as we breathe and our hearts beat without us having to consciously do anything. Tao points out that nature is full of instructive lessons and surprises. The weak can be strong, and the apparently strong weak. As the pioneering Taoist sage Lao-Tzu put it, 'nothing in the world is weaker than water, but it has no better in overcoming the hard'.

Tao, which offers psychological and philosophical insights rather than the structure of a traditional organized religion, started in China around 2,500 years ago – about the same time Buddhism started in northern India – and is in many senses the Chinese national religion. The followers of the way strive to nurture their *chi* (their intrinsic energy), to become one with nature through meditation and rituals, and to develop Tao's 'Three Jewels' of compassion, moderation, and humility. These were the aims pursued by a small community of monks for over 1,000 years in the Hanging Temple.

The mountain – like the one the temple hangs from – is the key to much Taoist philosophy and theology. The world – creation – is a duality composed of opposites that in Tao are called Yin and Yang. This concept represents the essence of all – night and day, negative and positive, male and female – and is embodied in the mountain. No mountain has only one side; it must have a shady northern Yin side in order to also have a sunny, southern Yang side. The point is that opposites go together – they define each other.

The temple is a dramatic affair. It hangs 50 metres (160 ft) high from the cliff-face, hovering beneath heaven and earth over a stretch of placid water that must once have been more fierce. The structure is audacious. It is built as one with the mountain – the two fuse together in an embrace. It's formed by a series of ornamental galleried pavilions, of different sizes and on different levels, linked by delicate timber walkways or by terraces fashioned out of natural crevices on the cliff-face. The main buildings – which were mostly rebuilt during the Ming dynasty in the sixteenth century – are lightweight structures sitting on a series of horizontal beams that are cantilevered out of sockets cut into the rock face. These beams carry most of the weight of the structures, but there are also a few stout vertical posts, built off levelled natural crevices, which provide extra strength. It's a fascinating and sophisticated structure – all its elements and details are related and, although ornamental, perform structural jobs.

In a sense, the whole structure is a demonstration of Taoist principles. All is working with, not against, nature – the design uses nature to tame nature. Here strength is achieved through apparent weakness; weight is not opposed by weight, as the structures are sustained not by massive vertical supports but by a cantilever system that, in a most elegant and minimal manner, utilizes the structural principles of nature. It's a system that employs opposing forces: the more the horizontal beams supporting the temple pavilions are forced down by the load they carry, the more firmly there are embedded in their sockets.

The arrangement of the 'hanging' pavilions and walkways is inspired by Chinese temple plans, but reorganized in response to the nature of the site. This is a temple like no other; walking along its narrow galleries feels like flying in the energizing air of the valley. This is sacred architecture with an almost visceral punch.

62

THE NEW SACRISTY
Basilica of San Lorenzo, Florence, Italy | 1520–34 |
Michelangelo Buonarroti

The miraculous Basilica of San Lorenzo, constructed in the first half of the fifteenth century to the designs of Filippo Brunelleschi, is only part of a remarkable complex of buildings. A small door at the altar end of the Basilica, leads into the New Sacristy. This room, one of the most magnificent creations of the Renaissance, was conceived and realized during the 1520s by Michelangelo to serve as a tomb and monument for the noble dead of the illustrious Medici family. It is the only building designed by Michelangelo to have been executed and largely completed under his supervision – and the only place to see sculpture, designed and at least partly carved by Michelangelo, still in the setting he created for it.

The chapel is square in its plan, upon which stands a cubic volume supporting a dome that rises from pendentives and with an inner surface that is coffered. Evidently the dome is a homage to the perennially inspirational Pantheon. The chapel's domed form is based on Filippo Brunelleschi's earlier Old Sacristy, with, in traditional manner, the cube representing the world of man, while the spherical dome is the world of God. So, symbolically, the design speaks of the soul ascending from the world below to the heavens above. This sense of apotheosis is reinforced by the architectural treatment of the interior. The four walls of the chapel are each like the elevation of a Roman triumphal arch, implying the heroic and transcendental status of the entombed Medici. The elevations get less visually busy as they rise from floor to dome – the lowest portions contain the tombs, while the upper are plain, with each containing only central windows, the sides of which taper from top to bottom, clearly implying upward movement. The message of this stripping away of detail is clear: the material world is gradually left behind with the ascent into the more abstract and simple world of the spirit, represented by this simplifying of detail that symbolizes the liberation of the spirits of the dead Medici as they flew to heaven.

Although the design of the New Sacristy is, in broad terms, inspired by Brunelleschi's Sacristy, it is clear that what Michelangelo had done was way beyond mere architectural imitation. Brunelleschi, and architects of the succeeding couple of generations such as Donato Bramante, had rediscovered and recreated Roman classical architecture. They had done so with gusto, verve, and imagination, but their essential goal had been to reconnect with an epic past, to emulate it and match it in quality. But Michelangelo wanted more. He wanted not just to master the past but to enrich it, to give the architectural language of Rome a new relevance. It was an intensely personal business. His classicism is radical and inventive – all is thrilling, fresh, and provocative, and like no classical architecture that had been produced before. Everything is the product of Michelangelo's fertile imagination. His key classical details and compositions, rendered in sepulchral white marble, challenge classical conventions – details flow one into the other, overlaying and overlapping.

And within this novel architecture sit the sculpted figures and the strange sarcophagi. The figures are extraordinary, and their meaning – even, in some cases, their authorship – remains debated. What is clear is that Michelangelo conceived and designed the main figures and the sarcophagi, with their characteristic curved tops, on which figures recline. It is also certain that he carved at least some of the figures, or parts of some of them, and that the sculptural scheme was left unfinished. This confusion and uncertainty is explained when the chronology of construction of the New Sacristy is placed within the context of Michelangelo's heavy workload and the political upheavals taking place in Florence in the 1520s.

By about 1524 the construction of the New Sacristy was complete and Michelangelo turned to the creation of the sculpture, but this attention was not undivided, for in 1524 he was also obliged to commence the design of the adjoining Medici library, named after Lorenzo the Magnificent. By this time Giulio de Medici had been elected as Pope Clement VII and the family's grip on power seemed certain.

The Medici to be honoured in the New Sacristy were Lorenzo the Magnificent and his brother Giuliano, together with a pair of relatively insignificant decedents named after them: Giuliano, Duke of Nemours, and Lorenzo, Duke of Urbino. Statues of the Duke of Nemours and of the Duke of Urbino were duly finished, and powerful seated figures, clad in Roman costume, sit brooding in the centre of the opposite walls.

But more compelling – and mysterious – are the four figures that Michelangelo created to lie on the tops of the sarcophagi placed at the feet of these two men. Below the Duke of Urbino are placed a male figure representing Dusk and a female figure representing Dawn. The female figure – naked and muscled like a man, with breasts that appear to have little to do with the body to which they are attached – wears a strange expression: languid, exhausted, more an image of death than of Dawn. The image of Dusk is a naked man whose face, turned to the spectator, is rough and barely finished.

Below the Duke of Nemours are also a male and a female figure. The male represents Day and, as with Dusk, his face seems to have been only roughed out. The female represents Night; she is the most complex of the New Sacristy's four emblematic figures and is generally agreed to be entirely the work of Michelangelo, not least because she leans on a mask which is evidently a self-portrait of the artist.

All these images have to do with time, that enemy of mortal man which marks remorselessly the path from birth to death, and when working on the New Sacristy death must have weighed greatly on Michelangelo's mind – not just the death of others, but also his own.

In the late 1520s and early 1530s, politics in Florence took a dramatic turn and the city became a most dangerous place. In 1527 Rome was sacked by the Holy Roman Emperor Charles V, and this led to the fall of the Medici and the establishment of a Republic. Michelangelo, although working for the Medici, actively supported the Republic, which put him in a very difficult position indeed when, in August 1530, the Medici retuned to power. Quite reasonably, Michelangelo believed his life was in danger – and no doubt for a time it was. He may have taken refuge within the New Sacristy, working half-heartedly on the figures, dreading the arrival of a Medici assassin. In this setting his self-portrait, lolled upon by the figure of Night, takes on a specific meaning – it's his own death mask, expressing his fear that Night had come upon him. But Michelangelo survived the wrath of the Medici, fleeing in 1534 to Rome, where, perhaps as atonement, the Medici Pope Clement VII, when at death's door, commissioned Michelangelo to paint a vast mural in the Sistine Chapel. The initial idea was that it should portray the Resurrection, but the next Pope changed the theme, slightly but subtly, to the Last Judgment – showing not just salvation but also punishment. Such was the sombre and menacing mood in the Vatican at the time of Martin Luther's 'heretical' Protestant Reformation.

63

IMAM MOSQUE
Isfahan, Iran | 1611 | Shaykh-i Bahai with Ostad Ali Akbar Esfahani

Shah Abbas the Great made Isfahan his capital in 1597, with the intention of establishing it as the greatest city in the Muslim world. He was the most able member of the Safavid dynasty that ruled the land from 1501 to 1722 and from 1729 to 1736, and his new capital was to be both a sacred city and a centre of trade. Shah Abbas encouraged cultural and commercial connections with Europe, and did much to develop a passion in the West for Persian silk and carpets.

The huge and architecturally handsome bazaar that Shah Abbas created still functions in central Isfahan, and still forms the material prelude to the city's stupendous central square, designed to evoke the beauty of a celestial garden of paradise.

The garden square – now known as Imam Khomeini Square, but originally called the Maidan-i-Shah or Royal Square – forms the heart of Shah Abbas' sublime city. It was started in 1611 and measures 500 × 160 metres (1,640 × 525 ft), and this proportion – roughly three to one – is significant because it is one of the key proportions of Solomon's Temple in Jerusalem, as described in the Old Testament. This reference was surely intended to suggest that Shah Abbas was the Solomon of his age and that his capital was a holy city.

The square is lined with sacred, royal, and commercial buildings – and so seems to suggest the world in miniature –

but it is dominated, on the side opposite the bazaar, by the mighty and architecturally dazzling Imam Mosque.

Built from 1611 for Shah Abbas, the mosque forms the tremendous architectural and spiritual climax of the journey from the bazaar and through the square. Indeed, the square, with its pools – a familiar Islamic image of earthly paradise – is like the forecourt in front of the Holy of Holies in Solomon's Temple. It is the prelude to the Imam Mosque – the journey across the square must have been intended to give worshippers the opportunity to detach themselves from the material world before entering the House of God.

The Imam Mosque, the great architectural glory of Isfahan, seems to have been the conception of Shaykh-i Bahai, a Lebanese-born genius who was a pioneering astronomer and mathematician as well as a theologian, philosopher, poet, and architect, and Shah Abbas' master-planner for Isfahan. Certainly Shaykh-i Bahai designed the setting for the mosque – notably the Maidan-i Shah – while the detailed design of the mosque itself was probably the responsibility of Ostad Ali Akbar Esfahani.

Like all mosques, the Imam Mosque is an echo of Mohammed's house in Medina, so, like traditional domestic architecture of the Middle East, it has an entrance gate leading to a central courtyard, off which are placed subsidiary rooms and courts. And, as with the houses of men, in this House of God water is used to cool the air, and interiors have lofty ceilings and screened windows to keep out direct sunlight but allow breezes to waft inside.

The entrance gate is stunning. Its surface is covered with polychromatic glazed tiles that sparkle and ripple in the sunlight, with the gate itself set within an *iwan*, a tall arched recess. The decoration on the tiles, if other than texts, must be geometric or abstract – inspired by the works of God as seen in nature, but never including renderings of living beings lest these be seen as idols challenging the fundamental Islamic creed that there is only one God, and nothing else must be worshipped.

Once across the threshold, the sacred geometry of the mosque becomes apparent. As with Judaism and Christianity, Islam believes the divine is expressed through mathematics and geometry, that numbers and proportional harmony are the building blocks of God's creation. The exterior elevation of the entrance gate is set at right-angles to the main axis of the square, but the moment the mosque is entered, the geometry shifts. The main axis of the mosque is diagonal to the axis of the square, and with good reason: the square is to do with man, while the mosque is to do with God. The exterior of the mosque relates to the works of man, while the interior is organized around the geometry of God. The diagonal axis followed by the main sanctuary of the mosque and the *mihrab*, or sacred niche, that it contains are aligned, ultimately, with the Ka'aba in Mecca and so proclaim the *qibla* – the direction in which prayers should be offered – to the faithful.

The threshold of the mosque is the point where the axes of man and of God meet; the moment the mosque is entered, the visitor orients towards Mecca. There is a powerful and simple beauty in this geometry.

The central court, leading to the sanctuary, has four *iwans* in the centre of its four sides. This arrangement is an evocation of paradise, which is described in the Old Testament and in the Koran as possessing four quarters. So this court is designed to read as a promise to the faithful of the afterlife that awaits them. In the centre of the court is a pool, which is not just ornamental but also functional. It helps cool the air and provides water with which worshippers can wash their hands, feet, and faces before they pray. In Islam, godliness and cleanliness go hand in hand.

Inside the main sanctuary, the cubical lower volume is topped by a huge dome. This is a traditional and familiar combination in which the dome, formed of a circle with no beginning and no end, symbolizes the spiritual world, while the cube represents the material world.

The dome over the main sanctuary is a structural wonder. Made of kiln-fired brick, it was completed in 1628 and is a major engineering feat that compares with the heroic domes of Europe – the slightly earlier St Peter's in Rome, designed by Michelangelo, and the slightly later St Paul's in London, designed by Wren. Like both these great Christian domes, symbolizing the heavens, this dome at Isfahan is multi-skinned, comprising an inner dome calculated to please the eye when viewed from below and, built off it, an outer dome to impress from afar.

Together, this magnificent mosque and the paradise square over which it presides form one of the greatest examples of the Islamic sacred city.

64

TAJ MAHAL
Agra, India | 1632–53

As one of the most famed historic sites in the world, the complex of structures that form the Taj Mahal at Agra is perhaps the last place on Earth where you would expect to be surprised by an architectural secret – but I was.

I arrived at the gates of the mausoleum very early in the morning as the first flickers of red-tinted sunlight touched the white marble of this most famed of sepulchres. In front of me was a garden, with its four rectangular slivers of canal evoking the four rivers of paradise mentioned in the Old Testament and the Koran; the script above the entrance gate that I'd just passed through proclaimed, 'enter thou my paradise'.

Beyond the garden is the focus of this creation – the domed tomb, raised on a platform and flanked by four minarets. And beyond this closed world of the tomb, this place of death and memory, is, just visible, the world of the living, the Yamuna River that on its more distant bank once sported Mughal structures and gardens of great significance.

The domed tomb, along with the rest of the complex, was built between 1632 and 1653 by the Mogul emperor Shah Jahan to serve as a sepulchre for his third wife, the beloved Mumtaz Mahal, and – arguably – also for himself. The tomb is a perfect piece of primal geometry, formed of two distinct parts that together occupy the volume of two cubes, one placed upon the other. Within the volume of the bottom cube sits the building proper. This takes the form of an irregular eight-sided figure in which four sides are long and four are short, with the long alternating with the short. Set in each of the four long sides is an *iwan*, or arched recess – a route of penetration.

Above this octagonal base, and occupying the volume of the upper cube, is a dome. But this is not just any dome – it is of that variety beloved by Islam, and looks, in form, like an onion. Why this shape? Why a dome? Islam must have been deeply impressed by the great domed early Christian churches – the Holy Sepulchre in Jerusalem and Hagia Sophia in Istanbul – and soon embraced and developed the form, with the Dome of the Rock in Jerusalem of the 690s being one of the earliest and best domed Islamic sacred buildings.

The symbolic meaning of the contrasting forms of dome and cube – representing the immaterial and material worlds – is familiar, so there were no surprises in this sacred geometry. Nor was the delicacy of the execution of the tombs really a revelation; even when seen from afar, the superb quality of the Taj Mahal is obvious. As I got nearer I could confirm how exquisite it is: white marble, finely cut, and embellished with twenty-eight types of precious and semi-precious stones. There is a jewel-like perfection about the structure, which feels like a mighty and rich casket rather than a building – a casket made to contain the emperor's greatest treasure.

I entered – and then came my surprise. Of course the tomb is dark and sombre, and so it should be, for it is to do with death. There is no direct light; instead, it filters in through a series of grilles. But what I didn't expect was the sound. This building for the dead is itself alive – alive with sound. It speaks, or rather groans. It's extraordinary. The sound is not the echoing reverberations that you might expect from a masonry-built, domed structure. It's much more unusual.

I walked around inside, past the graves of Mumtaz Mahal and Sha Jahan, and as I walked, the building sighed; it wailed. I suppose it was the wind, entering through the grilles and then rolling around the underside of the dome. I don't know, but the tomb inside certainly packed an unexpected emotional punch. The solemn light and the rhythmic sobbing and sighing of the stones was hypnotic. It's a structure that makes death tangible and the grief of Shah Jahan eternal.

65

TEMPLE EXPIATORI DE LA SAGRADA FAMILIA

Barcelona, Spain | 1882–1926 and 1952 to c.2026 | Antoni Gaudí

The Sagrada Família – the Expiatory Temple of the Holy Family – is without doubt one of the oddest major buildings of modern times. It is also, for some, one of the most unsettling. It weds concepts that appear to be almost in conflict, certainly in contrast. Such vast architectural projects have, by convention, to be programmed, planned, and detailed in coherent and consistent manner, with the design resolved from the start and related in a rational manner to means of construction, budget, and various other constraints. Yet in many ways this building is a monument to the spontaneous, the irrational, and the unplanned – or at least to the opportunistic.

It celebrates architecture as a great organic art, and buildings – the children of the imagination – as living and ever-evolving creations. And, of course, this spirit of life, of invention, of changefulness, can appear particularly appropriate for a great church, a house of the living God, whose construction should be not only an act of love, engaging all the senses, but also, possibly, an act of penitence.

The Sagrada Família is, almost inevitably, the result of a vision. In 1882 Josep Maria Bocabella i Verdaguer, while in prayer before an image of the Holy Family, received the 'call' to build a temple to expiate – or repair – the damage done by the sins of mankind. Many great cathedrals of the Middle Ages were places of pilgrimage; this was to be a place of atonement and of spiritual healing.

The temple was started in neo-Gothic style by Francesco de Paula del Villar, Antoni Gaudí's architectural master, but just as the crypt was being completed del Villar abandoned the project following disagreements with the builders and architect Joan Martorell. This display of all-too-human vanity or petulance was hardly an auspicious start for such an ethereal and sacred project, but as fate would have it, Gaudí was standing by. He was recommended by the building foreman and Martorell and on 3 November 1883 took over the job. Gaudí – pious, with a passion for nature and crafts and a mystic belief in the power and source of creativity – was just thirty years old and would stay with the job until his death in 1926, when the building was still only a quarter complete.

The design and construction of the Temple became increasingly personal and fantastical. Gaudí would claim it as an expression of Catalan identity, individuality, and artistic independence, and certainly many now would agree. The journey started as a fusion of Gothic Revival architecture of a most idiosyncratic type with a particularly organic form of Art Nouveau and a touch of Expressionism. Since these traditions are rooted in a veneration of natural forms, the combination promised – and has proven – to be most fruitful.

From the start Gaudí focused on the aspiring aspect of the design – on a cluster of spires to act as fingers to heaven, and on external façades calculated to carry profound spiritual messages (for example, the Nativity façade, with a Solomonic spiral column rising from the shell of a turtle, suggesting nature as the foundation of Christianity).

He also worked on the realization of a most extraordinary major internal space where Gothic-inspired nave columns and ribs and vaults are reimagined in the most bizarre manner to create an unprecedented interior – part parody of Gothic, part evocation of a slightly unhealthy looking nature. You either love it or you hate it, but you certainly can't fail to notice it, or Gaudí's tireless exuberance and invention.

The manner in which the temple has been wrought and embellished is a source of never-ending fascination. Details have been designed to appear as if they have not been designed but have simply sprouted in place; surfaces have been filled with ornament, suggesting a horror vacui on Gaudí's part; materials are mixed in extraordinary manner, and some chosen with what many have viewed as dubious taste – notably the rather kitsch-looking fragments of ceramic, set in unseemly cement. Most impressive, perhaps, are the stone-built external elevations that have been virtually sculpted, with figures and architectural and organic forms swirling together in mesmerizing manner, as on the Nativity façade.

The construction of the temple slowed due to lack of funding, stopped during the Spanish Civil War, and did not start again until 1952. Since then work has been spasmodic but almost continuous. The main challenge, of course, has been to realize such a personal design in the absence not only of the architect but also – in many cases – of any drawings (many were lost in a fire in 1936) and even any stated intentions. Inevitably the completion of the Temple has become a fascinating work of collaboration, a communal project undertaken by artists with an admiration for Gaudí who have been obliged to create in the spirit of the master, anticipating rather than realizing his unstated desires. For example, Gaudí left no very clear idea of how the scenes showing the Passion of Christ should be realized and arranged on the Passion façade. Since 1985 this project has been undertaken by sculptor Josep Maria Subirachs, who had incorporated Gaudiesque forms, and an image of the architect himself staring at a portrait of Christ in, it must be assumed, an expiatory manner. Currently work continues on what is, perhaps literally as well as spiritually, an eternal project.

66

GREAT MOSQUE
Djenné, Mali | c.1907

Djenné is a remarkable place. Anciently it was an important location of the great trade route through what is now Mali, along which goods, ideas, and religion travelled, transforming the land. Now it is a small provincial town that is notable because its buildings are low-rise and virtually all are of traditional mud construction. A walk through the town is a most memorable experience. The walls of the buildings are of undulating and organic form, made of sun-dried brick coated with a render that is a mix of mud, straw or rice husks, and dung. The buildings in the town include the Great Mosque, said to be the largest mud-built structure in the world still in use.

Mali is a fascinating land, once the heart of a great African empire that stretched to the west coast of the continent. Many of its inhabitants were converted to Islam by Muslim Arab missionaries during the thirteenth century, and by the fourteenth century Mali was a key part of the Muslim world, with strong trade connections to the north and east – connections maintained and developed by the waters of inland rivers, like the Niger and its tributaries, that formed natural routes of trade and communication. The empire fragmented until it was finally gobbled up, along with much of north and west Africa, by the French during the nineteenth century.

The Great Mosque, in its physical size and the awe it generates, seems emblematic of Mali's past greatness. It's a mighty and impressive affair – a double cube in shape, with a series of bastion-like towers that carry minarets, all well over 10 metres high. Its form is bold, abstract, elemental. It stands on a podium about 3 metres (10 ft) high that was probably originally reached via seven steps. The Koran says there are seven heavens with God above them all, and presumably this is what this podium and steps signify – the transition from the profane to the sacred. The walls of the mosque are smooth, but undulate slightly in an organic, almost sensuous way, and are pierced by rows of protruding palm tree trunks. Called *torons*, or horns, these act as permanent supports for scaffolding, for all mud buildings require frequent maintenance; with it they will last indefinitely, without it they will rapidly wash away and crumble in the rains. So, after each rainy season, and sometimes more often, workmen dangle from these poles and give the mosque a new mud-based skin.

The history of the Great Mosque is, in many ways, the history of the nation. The first mosque was built here in the thirteenth century by King Koy Konoboro, immediately after his conversion to Islam. During this period, Djenné – located on a wide and majestic tributary of the Niger called the River Bani – became one of the trading centres of the world, and in the region second only to Timbuktu, 500 kilometres (300 miles) to the north, as a centre of learning and pilgrimage. This was Djenné's heyday.

But what exists today is not King Koy Konoboro's creation, although it stands on its site. The original mosque fell victim to feuding between local Islamic sects, one of which believed the mosque defiled and demolished it in 1834. But in 1907 the current structure was built on the site of the original mosque, and although nobody suggests that it incorporates any early fabric, it is possible that its plan reflects that of its thirteenth-century predecessor.

The interior is gloomy and crowded with huge square piers. The double cube interior measures 50 × 26 metres (165 × 85 ft), and the number and close spacing of the fat-girthed piers is necessary because long and strong logs are not readily available in the region so the roof structure had to be formed with palm trees, and these are simply not capable of spanning a wide area.

These massive mud-built piers divide the volume of the mosque into a series of narrow aisles. There is no clearly defined central or major space, but a little emphasis is given to the aisle that focuses on the simple *mihrab*, the niche marking the direction of Mecca. The atmosphere in the mosque is remarkable – tranquil and decidedly cool. This is to do with the construction. Building with sun-dried clay brick is not only very economic and ecologically sound (no valuable resources or non-replaceable materials or sources of energy are consumed), but also very functional and appropriate in a hot climate. Thick earth walls with small windows and cross-ventilation keep the interior cool and comfortable during the day. At night, the heat stored in the walls during the day is released, rises, and if required is removed from the building by means of holes in the ceiling that can be easily opened or closed from the flat roof.

The Great Mosque is a spectacular example of architecture built from easily available local materials, and a vivid reminder that much vernacular building – non-destructive, non-polluting, and sustainable – has important lessons to teach.

67

CHURCH OF THE SACRED HEART
Vinohrady, Prague, Czech Republic | 1928–32 | Joze Plecnik

This remarkable church is one of those crucial links in the story of architecture that, in brilliant and illuminating fashion, connects tradition, technology, and culture. First impressions are vital: it is evidently a church, and its stylistic affiliations are clearly classical, but it has about it the feel of industrial plant – a power station of religion. And its materials of construction are unexpected for a building designed in the classical tradition. The lower portion of the body of the church and of the tower are wrought, or at least clad, in brick, a traditional building material, but with a crisp and clipped industrial quality. However, the upper potions of both body and tower are formed of a very different material, a pure, seamless white render that has the appearance of cast concrete. The details and forms are idiosyncratic and abstract; most notable is the detailing of the brickwork, where the conventional Flemish bond is studded with projecting vertical slabs of stone arranged in an almost hypnotically close-set and rhythmic pattern that casts extraordinary shadows in the raking sunlight. The doors, with their rendered portals that flare upwards and outwards, seem to be inspired by the seventeenth-century Baroque of Borromini but are said to be an evocation of a priest's robes and influenced by late nineteenth-century experiments with textile references in architecture.

But most distinct and visually powerful is the tower, framed by two slender pyramidal obelisks – traditional emblems of death and resurrection – and pierced by huge round windows containing clock faces. Wide but narrow, the tower is calculated, in a most sculptural manner, to transform the character of the church when perceived from different angles. Seen head-on from the west, the tower makes the church appear massively broad and squat, with a wide and shallow lower pediment echoed by the higher, narrower, but equally shallow pediment that tops the tower. But side views suggest a very different volume, with the tower reading as no more than a sliver of a campanile. This manipulation of form for emotional effect and to transform the perception of a building must remind an English observer of the early eighteenth-century Baroque churches of Nicholas Hawksmoor, particularly Christ Church Spitalfields, with its wide but shallow tower that works visually in an almost identical manner. Could Joze Plecnik, the architect of the Sacred Heart, have known this building? Almost certainly.

If the external appearance of Plecnik's tower is thought-provoking, its interior is little short of miraculous. Within its long but narrow bulk, and across its pair of wide round windows, rises a concrete ramp of switch-back form. Swathed in light, it is an image of industry and of Modernism reminiscent of the *promenade architecturale* that Le Corbusier had used in the slightly earlier Villa Roche and in the contemporary Villa Savoye.

Plecnik was born in Ljubljana in 1872 and from 1895 studied architecture in Otto Wagner's studio in the Viennese Fine Art Academy. In 1901 he was appointed secretary of the Secessionist movement. Like Wagner, Plecnik was on a quest to forge a modern architecture based on tradition and to evolve a new and distinct decorative style. It is tempting to see the stone studs in the brick elevations of the Sacred Heart as a permutation of the stone 'bolts' that Wagner used earlier (1904–06) on his Post Office Savings Bank building in Vienna.

But for Plecnik, architecture had a very special dimension – he wanted to use it to express Slovene and Czech pride and culture within the rambling Austro-Hungarian Empire. By 1911 Plecnik was based in Prague, teaching and practicing architecture, and, following the creation of an independent Czechoslovakia in 1918, in the wake of Austrian defeat in the First World War, he started to evolve an architectural language to express the new nation's identity. In common with other early twentieth-century theorists, Plecnik resolved that the early peoples of the new nation had Aeolian and Ionic roots, and so the use of primitive Greek classical forms

– particularly the Ionic order – became for Plecnik a statement of national identity. This conviction can be seen in the splendidly vigorous and inventive classical additions he made within Hradcany Castle in Prague – the presidential residence – from 1920, notably the extraordinary hall of 1927, with its tiers of columns, and the Bull Gate of 1929, with its witty Ionic canopy; and in the Doric and Ionic detailing of the staircase in the University Library in Ljubljana, from 1936.

The interior of the church reflects Plecnik's commitment to classicism as a national style. The context of the details is quasi-industrial, in particular the brick walls articulated with plain pilaster strips and a gallery like a factory walkway. But this utilitarianism serves in wonderful manner to highlight the fine quality of the marble altars, with their inventive classical details. The crypt is perhaps the most satisfying space – a semi-circular barrel vault of brick, reminiscent of a Roman bathhouse, with an exquisite columned altar. This is elemental architecture of the very best sort, demonstrating the eternal truth of the observation, used by Mies van der Rohe for a very different architecture, that always 'less is more'.

68

NOTRE DAME DU HAUT
Ronchamp, France | 1950–54 | Le Corbusier

What would a man like Le Corbusier produce when designing a pilgrimage chapel for Roman Catholic worship? He was not known for his Catholic convictions or Christian religious faith – indeed, as an 'initiate' and often misunderstood champion of Modernism, he viewed himself as a heretic. Some have fanaticized that he was emotionally wedded to the anti-materialist beliefs of the Cathars of the south of France, the people to whom the Swiss-born Le Corbusier traced his roots and who in the thirteenth century were regarded by northern French Catholics as heretics. It is naive to suppose that an architect has to be of a particular faith before he can design a building for its use – but surely it helps. Certainly, Le Corbusier showed little enthusiasm for the commission when it was first suggested. He was then locked into the problem of forging a contemporary architecture to help solve the material ills of the modern world – to humanize dark, polluted, and war-torn cities, and, as with his Unité d'Habitation scheme for Marseille, to provide higher-quality public housing, at speed and in volume. He was finalizing the Marseille project when he was asked to design the chapel at Ronchamp.

He visited the site to brood and seems, after a few days, to have been gripped by a vision. Clearly the potential relationship between a proposed building and the hill-top site seized his imagination, as did the rich potential of sacred architecture, evoking multiple ancient images, geometry, mathematics, and systems of harmonic proportion that by tradition were seen as the building blocks of divine beauty and creation. And, it seems, as in a dream, he perceived the artistic potential of the dualistic nature of most religions – life and death, light and shade – and the essential Christian and Cathar notion that God dwells not in the material world but in the immaterial, in light and in colour, in the space defined by structure. This religious duality of course reflects Le Corbusier's own dualist architectural 'faith', expressed in his

attempt, as he had explained in 1939, 'to resolve the dichot-
omy between the Engineer's Aesthetic and Architecture, to
inform utility with the hierarchy of myth'.[5] Le Corbusier was
so much more than a mere doctrinaire Modernist believing
that form must follow function – he was also an artist, and as
this chapel demonstrates, a master of the mystic and the
seemingly irrational.

After a few weeks of gestation, Le Corbusier grabbed a
piece of charcoal and, entranced in a moment of revelation,
he rapidly sketched his initial idea for the building. As Robert
Coombs asks in his brilliant book *Mystical Themes in Le
Corbusier's Architecture in the Chapel Notre-Dame-du-Haut at
Ronchamp*, was he suddenly 'transported by some passing
muse'?[6] More certainly, as Coombs points out, the chapel
contains forms, proportions, and details that relate to much
sacred architecture and that make it, perhaps, not simply a
Roman Catholic chapel but – more dramatically – an almost
universal, elemental sacred building. In a marvellous and
almost mystic way the chapel acts as a threshold between the
material and spiritual worlds; it is almost an incubator for
reflections upon the divine, and is a perfect place for religious
contemplation.

Coombs notes that in his design Le Corbusier evokes
images associated with the Virgin, and that he pursues an
'alchemical' theme of conflict between 'cosmic forces' and
transmutation from base to fine, and a 'Catharist eulogistic
theme …possibly shaded with Knights Templar philosophical
connotations'.[7] It is probable that the spatial arrangement of
the chapel echoes a thirteenth-century 'shrine' of Montségur
where Catholics martyred Cathars and thus is evidence,
observes Coombs, of 'architectural sleight-of-hand', with Le

Corbusier using one shrine, that of the Catholic victors, to 'slyly' commemorate another, that of the Cathar martyrs.[8]

So Ronchamp is, arguably but demonstrably, a complex interplay of Marian, alchemical, and Cathar themes, spiced with astrological references, and all, as Coombs rightly states, affirmed through mathematics and 'governed by geometrical regulators',[9] with the Golden Section and the *vesica piscis* – the traditional geometric emblem of the Virgin – dictating the design of the chapel's plan and south door.

How on earth did Le Corbusier unite all these themes to create a poetic, powerful, and coherent work of architecture? Through art, and by what Coombs calls his 'Cubist methodology', Le Corbusier's artistry is apparent everywhere. There is the art of architecture, naturally, expressed most obviously in the extraordinary forms of the building that seem at once inspired by nature but also intensely and consciously sculptural; these forms are evidence of Le Corbusier's conviction that strong emotion can be extracted from sublime and elemental abstract composition. His artistic sensibility is also revealed in the choice of intense colours and their juxtaposition with rough-textured white walling; in the astonishing and collage-like organization of the internal fenestration; in the design of the stained glass, often set in deep window reveals, suggesting that the building is a sacred cavern; and in the chapel's organic interior, which, combined with the red hues from the coloured glass can, perhaps with the help of an overheated imagination, be seen to symbolize the sacred womb of the Virgin, the quickening chamber of the Saviour of the World. And, of course, it can be seen in the 'paintings' by Le Corbusier, most notably the enigmatic south door, rendered in enamelled metal, which includes ancient sacred symbols and forms: the pentagram, the pyramid, and the coiled serpent, perhaps representing the latent human energy released through Kundalini yoga. A universal church indeed.

CHAPTER 4

URBAN VISIONS

THE CITY IS ONE of the great benchmarks of civilization. Indeed, the very word 'civilization', originating from the Latin *civis*, meaning citizen, is rooted in urban virtues. The first cities, emerging perhaps between 5,000 and 6,000 years ago, brought security, shared enterprise, amenity, and industry, which created wealth, trade, culture, and law – essentially those qualities that are emblems of civilization. Cities also had a huge architectural consequence, giving rise to ideas about planning and zoning; the segregation and inter-relationship of uses; the integration of architecture and gardens, and the evolution of the principles of water supply and communal living. Cities also gave birth to a series of significant building types, ranging from city defences to urban temples, palaces, and markets, and to shops and houses of coherent design, such as those that form uniform terraces, arcades, and gallerias. At moments in world history cities have been expressions of human endeavour, architectural ambition, and creative excellence. But cities can also tell a dark story, for they have fallen victim to destructive ideologies, religious intolerance, warfare, and lust for power. In many ways the history of cities is the history not only of architecture, but also of humankind.

This chapter looks at the earliest cities, now thought to be those founded in Mesopotamia (in present-day Iraq), and then at some of the great classical cities from 2,000 to 2,500 years ago, including the imperial Roman trading city of Leptis Magna in Libya and the sensational Arab cities that were inspired by the culture of Greece and Rome and which evolved into phenomenal places of urban and architectural excellence, notably Petra in Jordan and Palmyra in Syria. The ferocious events of early 2015, when people, architecture, art, and history were targeted by the self-proclaimed Islamic State (IS), a rogue power that practices cultural terrorism, means that some of the great and ancient cities included in this book – notably Palmyra and Hatra – have been damaged or are currently under grave threat. At the time of going to press Palmyra is occupied by Islamic State and is at its mercy.

Maya and Inca cities in Mexico and Peru are described, as are individual and highly influential urban buildings and compositions, notably the early seventeenth century Place des Vosges in Paris that established – with great gusto and success – the principle of designing separate buildings as uniform, palatial, and classical compositions. The Circus in Bath, England, conceived about 150 years later, is in several key respects one of the progeny of the Place des Vosges – although with many distinct and curious permutations.

The twentieth-century adventure in urban design and city-making, which saw the creation of politically and architecturally utopian cities such as Brasília in Brazil and the evolution of distinct urban building types such as the skyscraper, is touched upon in the chapter on Pioneers. In this chapter, the twentieth century is represented by the Sydney Opera House in Australia, which, after a turbulent process of design and construction, is now generally accepted as a building of great beauty and a symbol not only of a city, but of a whole continent.

69

URUK
Iraq | 3,200 BCE

Uruk's history is lost in myth and speculation, although there is now a general consensus that 5,200 years ago Uruk was the largest settlement in southern Mesopotamia and therefore almost certainly in the world, and so the world's first great city. At what was arguably the peak of its growth and architectural and social sophistication, around 4,900 years ago, Uruk was walled, covered an area of 6 square kilometres (2 sq miles), incorporated a series of canals connected with the Euphrates, was beautified by gardens, and had an estimated population of 65,000.

By this time Uruk also possessed other familiar characteristics of great cities, notably monumental buildings, both sacred and profane, wrought of sun-dried bricks and largely faced with kiln-fired bricks and clay cones that offered an opportunity of ornament and that protected the sun-dried bricks from the weather. At whatever date the Old Testament was written –

probably between 3,000 and 4,000 years ago – Uruk already had a reputation for antiquity and grandeur. It is almost certainly mentioned in the Book of Genesis, where it is called Erech and appears in the description of the progeny of the sons of Noah, including Nimrod, whose kingdom began with 'Babel and Erech ... in the land of Shinar' (Genesis 10: 9–10).

There are other ancient city-like settlements in the world. Jericho, now in the Israeli-occupied territory of the West Bank, could have an origin as a small oasis settlement dating back 10,000 years, while Mohenjo-daro, near Sindh in Pakistan, was built from around 4,600 years ago, in a most orderly urban manner. Harappa, in the Punjab, Pakistan, is perhaps a little more recent. Both Mohenjo-daro and Harappa were built of kiln-fired and sun-dried bricks, both display sophisticated civil engineering, and both were part of what is now known as the 'Indus valley civilization'. Harappa became the larger of the cities and 4,000 years ago had a population of perhaps 23,000.

But Uruk is of fundamental importance not just because it was earlier and larger than other early cities, but because Mesopotamia – the fertile region between the Tigris and Euphrates rivers – became the 'cradle of civilization' and the centre of Sumerian culture, which pioneered many of the things that we now regard as key emblems of human development. These include writing (cuneiform), originating around 5,200 years ago; the measuring and division of time by means of 60-second minutes, 60-minute hours, and 24-hour days, divided equally into daytime and night-time; a system of numbers incorporating the 360-degree circle that formed the basis of geometry; irrigation and the harnessing of wind power; craft technologies and, most importantly, the wheel; and of course the city and urban living.

Uruk was divided into eight districts, each with a different character. The central Eanna district, developed from about 5,400 years ago, contained temples and public buildings, while other districts contained courtyard houses – made of sun-dried bricks and designed to achieve high insulation and to utilize natural ventilation – that provided the model for house construction in the region for centuries to follow. Generally the houses in these residential districts were occupied by people in the same profession or trade.

Uruk's most famous character is Gilgamesh, by repute its king around 4,700 years ago and the subject of what is regarded as the world's oldest book, *The Epic of Gilgamesh*. Gilgamesh sought immortality, and found it ultimately through architecture and the construction of cities wrought in strong kiln-fired bricks. He realized that stamping his name on hard bricks, 'where the names of famous men are written', would mean that his creations, and the memory of him, would last for eternity.

The power of the kiln-fired brick as a passport to immortality was made clear as I walked over the now desolate ruins of the great city of Uruk. Everywhere lay ancient kiln-fired bricks – strong as the day they were made – and even sun-dried bricks, marking the remains of both courtyard houses and public buildings. Gilgamesh was correct: these bricks assured that the past remained alive.

The embedding of names on kiln-fired bricks became a standard procedure in the ancient world. At the 2,575-year-old walls that survive in Babylon, I saw many bricks bearing the name of their creator – Nebuchadnezzar II, the 'king of kings' – and I also found his name on scattered bricks in the Babylonian city of Borsippa, Iraq, organized around a mighty ziggurat, the brickwork of its upper storey fused by an ancient lightning strike.

And among the bricks in Uruk lay, in even greater numbers, kiln-fired clay and stone cones, and occasionally a small stub of mud-brick wall faced and weather-protected with cones. And when the dust was washed off one of these the walls, perhaps 5,000 years old, an enigmatic pattern emerged – chevrons, lozenge shapes, and spirals. These are the patterns found on much later mosques and mausoleums, such as that of Isma'il Samani, in Bukhara, Uzbekistan; and in Romanesque Christian churches, such as Durham Cathedral. But here, in the ruins of Uruk, I saw the origin of the mysterious and ancient sacred language of abstract forms.

The most striking surviving single structure in Uruk is the vast, 4,100-year-old Eanna ziggurat. Only its core of sun-dried bricks remains – its outer skin of kiln-fired bricks was stolen centuries ago – so it's now no more than a mighty mound, but its power and presence are enormous.

To walk over the ruins of Uruk, the world's first great city, is an unrivalled experience. What was once a vision of beauty is now one of heartbreaking decay and desolation. As the goddess Ishtar laments in *The Epic of Gilgamesh*: 'Alas, the days of old are turned to dust.'

70

PERSEPOLIS
Iran | 5th and 6th centuries BCE

The construction of Persepolis started in around 515 BCE, on a site almost certainly chosen by the first Achaemenid ruler, Cyrus the Great. But the major sacred, ceremonial, and palace buildings were constructed by Darius the Great, who came to the throne in 522 BCE, and by his son Xerxes I, who reigned from 486 to 465 BCE. By the mid-fifth century BCE, Persepolis was the heart of what was then arguably the greatest empire on earth.

The political potency and architectural ambition of Persepolis must have been clear to all who beheld it in its golden age. It was, as its name made clear, the City of the Persians, and was intended to make a powerful statement about kingship. The city, with its main buildings elevated on an acropolis, was to commemorate and celebrate Persian achievements and to serve the king as a comfortable and elegant summer capital, notable not for its fortifications but for its palaces, fountains, orchards, and formal gardens. The creation of idyllic, well-watered gardens to evoke a notion of a divine paradise echoes Old Testament imagery and seems to be of Persian origin, though it was later embraced by Christians and by Islam, with its paradise gardens symbolizing Eden.

Persepolis was indeed to be a paradise – a tribute to the world order that Persian power had created and, more particularly, to Darius. Indeed, Persepolis was a key part of a plan to sustain Darius in power. He had grabbed the throne in 522 BCE following a confusing tussle after the death in 530 BCE of Cyrus the Great and the subsequent deaths of Cyrus' two sons, Cambyses and Bardiya (the latter was almost certainly murdered).

Whatever Darius' involvement in these deaths, he, a distant cousin of the dead sons, was the beneficiary. After this violent interlude Darius revealed himself to be a man not so much of action but of administration. He failed in his attempt to conquer all of Greece – stopped famously by the Athenians at the Battle of Marathon in 490 BCE – but he was determined to hold onto what he had seized and to establish his legitimacy.

So Persepolis was not a city of commerce or trade, nor really of military triumph, but one of political power and prestige, intended to display Darius' qualities and establish his reputation.

Persepolis, conceived and erected as a showpiece, must have had the feel of a model city, all its buildings well-considered, related, and jewel-like. Even now the ruins possess a sense of harmony that in the city's prime must have characterized its various buildings and their relationship to each other. The sense of removal from the ordinary world would have been emphasized by the fact that the city was built on a massive man-made terrace or plateau roughly 15 metres (50 ft) high, raising it above the rough terrain around it. For a city, Persepolis looks surprisingly small. That's because what we now see is really a group of palaces and related buildings set on an acropolis – spread around the retaining wall of the plateau would have been houses and shops, now long lost.

The main route of entry up to the royal and sacred plateau is disconcerting. Rather than a strong gate, or defences in depth, there is a pair of broad staircases leading away from each other, each turning 180 degrees at a half-landing and then running towards the other to meet at the top of the plateau. There were fortifications, but these embraced the larger city. All coming to the palaces on the plateau would have used this route. It was the start of an extraordinary journey that, 2,500 years ago, led to the epicentre of world power.

The staircase was calculated to impress and intimidate, for many of the people using it and the route to which it led

47 *The Hagia Sophia at sunset. This great sacred mountain of a building – vast and elemental – is dominated by its stupendous sixth-century dome. The minarets were added in the late fifteenth and sixteenth centuries after the church had been converted into a mosque. (Mehmet Ozcan/istockphoto)*

48 *The exterior of the Dome of the Rock, showing three of its eight sides – each embellished with seven-arched recesses, and decorated with tiles depicting geometric patterns and religious texts. Above all hovers the gilded dome, making this one of the most perfect and beautiful – as well as most sacred and mystic – buildings on earth. (Sean Pavone/Alamy)*

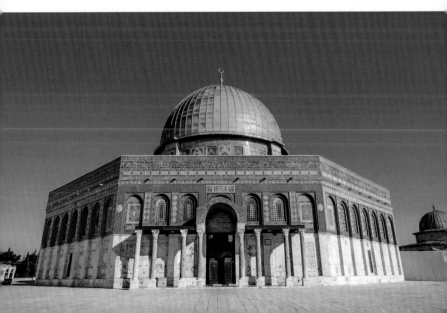

46 *St Catherine's Monastery, built in its present form in the early sixth century, looks like a small, well fortified town – which is not surprising since it doubled as a frontier fortress in a potentially hostile land. (Dan Cruickshank)*

49 *The spiral minaret of Malwiya Mosque in Samarra, Iraq, is a perfect, visually simple geometric structure whose spiralling ramp seems to suggest a journey to heaven. (yuliang11/istockphoto)*

53 *Chartres Cathedral, completed in its existing form by the late 1220s, is one of the first and greatest Gothic cathedrals in the world. Here, the potential of a system of ribs, piers, vaults, and flying buttresses – all utilizing the inherent structural advantages of the pointed arch – was fully realized. (Roland Baumann / istockphoto)*

54 *The Minaret of Jam is one of the most beautiful early minarets in the world. It is also now one of the most threatened. Dating from 1174 and built of hard, kiln-fired brick, its outer surface is embellished with geometric patterns, sacred symbols, and Koranic texts in Kufic lettering. (Dan Cruickshank)*

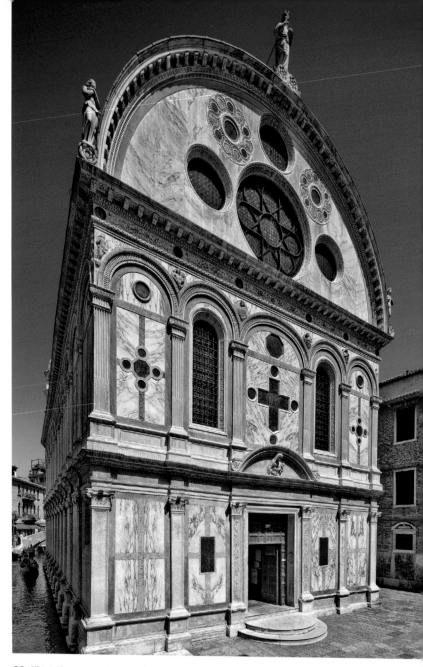

60 This is the entrance elevation of *Santa Maria dei Miracoli, a jewel of a building, clad with marble and ornamented with two tiers of pilasters. (Agencja Fotograficzna Caro/Alamy)*

61 *The Hanging Temple, formed by pavilions linked by walkways and incorporating small terraces, hangs from the mountainside. The structure seems to demonstrate Taoist principles – all its elements work in harmony with natural forces, utilizing a cantilever system that reflects the constructional principles of nature. The vertical posts are mostly for ornament and to reassure mere mortals that the structure is safe. (Dan Cruickshank)*

64 *The Taj Mahal at Agra, India, is one of the most famous buildings in the world and – still – one of the most satisfying to visit. Built between 1632 and 1653 as a tomb and shrine to the dead, the building is – through the beauty of its design, construction, and setting – a monument to the wonder of life and to the transcendental power of architecture. (Dan Cruickshank)*

68 *Notre Dame du Haut is here seen from the southeast. Its form seems to echo, in the most original and creative manner, the earlier chapel on the site, the thirteenth-century Cathar 'shrine'*

of Montségur, and the images and geometry traditionally associated with the Virgin Mary. (Dan Cruickshank)

66 *The Great Mosque at Djenné, Mali, possesses a marvellous abstract quality. It is probably the largest mud-brick structure in the world that is still in daily use. (Michel Piccaya / istockphoto)*

72 *The Arch of Septimius Severus, at Leptis Magna, Libya, is one of the most striking architectural monuments of this great ruined Roman city. The four-faced 'tetrapylon', built between 194 and 202 CE, marks the crossing place of the principal streets of the city. Above the arched opening are figures of winged victories. (Dan Cruickshank)*

71 *The Khazneh, or Treasury, at Petra dates from the first century* CE. *This sensational piece of sculpted classical architecture served as a tomb, including a triclinium, in which the living could feast with the spirits of the dead. (Holger Mette/istockphoto)*

73 *Palmyra, Syria, is seen here (top) in an image taken in 2007. Colonnade Street – the main spine of the 2,000-year-old city – runs from the Temple of Bel in the east (top) to the valley of the tombs, in the west. In May 2015 the city fell under the barbaric control of the self-proclaimed Islamic State (IS) During the time IS occupied Palmyra it severely damaged the the city's most significant structures including the Temple of Bel, the Temple of Baal Shamin and the*

Monumental Arch (shown before, left, and after, right). In most cases stones were tumbled by explosions but not utterly destroyed so authentic reconstructions are possible. Palmyra is one of the most important ancient city sites of the classical world, and arguably the most beautiful. Its future protection and careful repair are of the greatest importance. (Dan Cruickshank)

81 *The vast vault of the Taq-i Kisra at Ctesiphon, Iraq, probably dates from the sixth century. It remains the widest un-reinforced brick-built arch in the world. Here, scale was used two convey the power and grandeur of the princes who occupied such a bold, palatial structure. (Henry Arvidsson/Alamy)*

85 *Tower houses are grouped to form an outer perimeter of the city of Shibam. A house on the left – built of sun-dried wafer-like bricks – has not been regularly maintained and has crumbled, its brick gradually dissolving in the rains. (Dan Cruickshank)*

90 *The south or garden elevation of Marshcourt. The house is clad in blocks of chalk, excavated from the site. The large, mullioned windows on the left light the 'great hall' and the drawing room. The large block on the right, partly screened by trees, is the ballroom added in 1924–26 to Lutyens's design. (Dan Cruickshank)*

92 *This view of the Farnsworth House shows the main living area and, in the foreground, the entrance deck or terrace. (Dan Cruickshank)*

100 *The Temple of the Sun, or of Shamesh, at Hatra, perhaps dates from the first century BCE. Below, carvings of dromedaries and of the 'Green Man of Hatra' suggest a time when the people of the region venerated nature. These photos were taken in 2002 and sadly the Green Man has since been destroyed by IS. The future of the site remains in dire jeopardy. (Dan Cruickshank)*

were under the power of Persia. They were the conquered, and they came here to pay tribute and homage to their overlord, the king of the Persians or the 'king of kings' – a man who had the power of life and death over them, perhaps over their nations. It must have been a terrifying experience, with all the architecture designed to increase the supplicant's sense of awe and powerlessness.

At the top of the staircase are the substantial remains of the great cubical gate house built by Xerxes I, known as the Gate of all Nations, for all the nations of the known world had to enter here for audience with the Persian king. Within the gate, guarded by Assyrian-style human-headed bulls, is an inscription calculated to awe and humble all who read it. It declares: 'I am Xerxes, great King, King of Kings, King of Lands, King of many races . . . Son of Darius the King ... Many other beautiful things were constructed in Persia ... I constructed them and my father constructed them.'

From this gate, the ceremonial route splits in two. One path leads along a processional way – daunting in its scale and originally covered – to the throne in the Hall of the Hundred Columns, built by Xerxes. The other leads to the double staircase leading up to the Apadana or Audience Hall. This hall was, at the time, the most extraordinary structure in the world – a space built to amaze and daunt. It had a huge cedarwood roof supported on 20-metre (65-ft) columns, and its shady and mystical volume could accommodate up to 10,000 people.

The staircase up to the Apadana makes all explicit. It is embellished with beautifully carved figures that offer a microcosm of the Persian Empire. Those who used the stairs are commemorated on them, in startling and well-preserved bas-reliefs. Here it's possible to meet the people of the twenty-three subject nations ruled by Darius and Xerxes, and to see them bringing tribute to the Persian court. All the figures appear calm and peaceful – they are portrayed as having accepted their lot, and each group of tribute-bearers is led, in paternal manner, by a long-robed Persian official.

Other figures abound. There are Persians in their long robes and feather headdresses, and soldiers – the 'Immortals' – standing guard; and among these portrayals of worldly wealth and power there are also religious images. There is a ring with wings, which is the symbol of Akhura Mazda, the god of the Persian state religion, Zoroastrianism.

This use of sophisticated art to carry a political message is ancient, but the images – almost as fresh as new after being buried for centuries – still possess the power to shock.

71

PETRA

Jordan: the Treasury | c.100 BCE to 100 CE

Petra reached the peak of its power, size, wealth, and architectural beauty around 2,000 years ago, when it was the great trading city of the Nabateans, an Arab people of immense cultural sophistication. The city is cloaked in mystery and romance. It was only rediscovered by the West in 1812 after being largely abandoned during the sixth century CE, its ruins pillaged and its graves robbed. Since then, its strange location – secreted within a lunar landscape of mountains and reached via a long, deep, and twisting gorge called the Siq – has given Petra a compelling romance. And then there is the material from which the city is made. The word *petra* means rock, and the rock from which the city is wrought – for its architecture is carved rather than constructed – is a hauntingly beautiful, red-hued sandstone containing marble-like swirls of colour. Added to this is the quality of the surviving architecture, with finely detailed and vigorously inventive classical compositions cut into the faces of the cliffs that frame the city.

Petra is, in many senses, the great city of the dead. Not only has this once-thriving city been long abandoned – in 100 BCE it had a population of around 30,000 and controlled the valuable trade route between Arabia and Damascus – but its people were deeply in thrall to a death cult that led to the construction of Petra's most enduring monuments. These are thousands of rock-cut tombs, some of spectacular grandeur, that rise above and surround the site: the city of the dead standing guardian over the city of the living.

The Nabateans had occupied the site from the seventh century BCE and had a religion that was influenced by those of the great and ancient civilizations of the region: Egypt, Assyria, and Babylon. They had a pantheon that incorporated certain Greek gods, but at its core was their own Holy Trinity, as with the Arab people at contemporary Hatra in Mesopotamia. There was Dushara, the father and main god, the great protector, associated with Jupiter and with the mountains and high places. Then there was Allat, the goddess of trade, associated with Athena; and al-Uzza, the goddess of rain, fertility, and nature, associated with Aphrodite, Isis, and Ishtar. Mountains were holy – the gods were believed to occupy high places – so by living here and by hewing their mountainside tombs, the Nabateans believed that they and their dead lived with the gods.

The most famous of these tombs is knows as the Khazneh, or Treasury – an absurd name given to it centuries ago by the Bedouin, who fancied that every old structure or tomb must contain a secret hoard of treasure. In all probability it was a tomb or a *triclinium*, a dining room related to a tomb in which feasts were held as the living communed with the dead.

The Khazneh – more a sculpture than a building, for it was carved out of the living, sacred rock – dates from the first century BCE to the first century CE, with its quality, size, and location in the heart of Petra suggesting that it was associated with King Aretas IV, the man who presided over the golden age of Petra as an independent trading kingdom.

The elevation is erudite and rather flamboyant, in the free and inventive spirit of Roman provincial architecture. There is no question that, although the city was still politically independent of Rome, the arts of Petra were heavily influenced by the mighty brooding presence of the great empire. The tomb has a temple front formed by a lower half consisting of an entablature supported by six composite columns, and that in turn supports a pediment that stretches over the central four columns. Above this columned and pedimented façade sits – rather daringly, inventively, and most unconventionally – a second temple front, but very unlike that below. This also incorporates a colonnade, but each end pair of columns supports only a triangular corner of broken pediment, while the centre pair swirls around and reproduces to form a colonnade around a centrally placed round temple, or *tholos*, that is a perfect little self-contained building in its own right.

It is all very charming and elegant – the classical language of Greece adapted and reinvigorated in a playful manner. But it is the detail that is most revealing. Most of the carved images on the building are of goddesses or legendary women. For example, there are dancing Amazons and winged victories, and at the top, in pride of place at the centre of the *tholos*, is al-Uzza, in the guise of Isis. And the elevation faces east, towards the rising sun, in the direction of renewal and rebirth. All considered, it is reasonable to conclude that this is not only a tomb and *triclinium* (despite the huge and rather funereal urn perched on top of the *tholos*) but also a temple to female fertility, to regeneration and rebirth, to life following death as, each day, the sun sets only to rise again.

Indeed, as recent research has revealed, this elevation appears much concerned with time and the cycle of events. In fact, it can be read as a calendar, and of course a calendar has much to do with time and with fertility – there is a time to reap, and a time to sow. At the top of the elevation are four eagles, long supposed to be an act of deference to Rome – but eagles are also symbols of the sun, of the two elements of air and of fire, and here perhaps represent the four seasons of the year. Within the frieze of the lower entablature are seven cups, vessels for water but also representing the days of the week, and the cornice above the central door leading to the rock-cut chamber incorporates 365 dentils. Buildings incorporating key numbers from the calendar had been built before, and the 365-day calendar was also far from new when this elevation was carved. This was perhaps a tomb and a temple, but also surely a place of rebirth, where the body and the soul were to gain new life.

72

LEPTIS MAGNA

Libya: the Basilica of Septimius Severus | c. 200–215 CE

Leptis Magna, one of the greatest and most evocative ruined cities from the classical world, lies on the Mediterranean coast of Libya. In the tenth century BCE it was a Phoenician settlement, and as early as the sixth century BCE a great trading centre under the control of Carthage. But the city enjoyed its golden age in the early third century CE, during the reign of Emperor Septimius Severus, when the population neared 80,000 and it was the great gate through which commodities were traded between Africa and Europe.

The central part of the city – notably its public buildings – was excavated during the early twentieth century when Libya became part of Italy's empire. The discoveries were magnificent, but work stopped after Italy's defeat in the Second World War and little was done in the following decades, beyond a few isolated discoveries including the very high-quality 'gladiator' mosaics unearthed in 2005 and the general preservation and presentation of the ruins. The fighting that flared in Libya in 2011, and that continues at the time of writing as part of a bitter civil war, has caused damage and made the ruins insecure and their future uncertain.

Leptis Magna had been a possession of Rome since the early second century BCE, with the status and numerous privileges of an official Roman colony within the province of Africa. But Septimius Severus, who reigned as Emperor from

193 to 211 CE, was a native of the city, and he did all he could to make his home town great.

A most dramatic surviving token of Septimius Severus' favour is the splendid (although much restored) arch, bearing his image in the guise of Jupiter, that he built to command a key crossroads in the city. Through this curious cubical arch passes the Decumanus Maximus – the main road of the city – and the Cardo, once the main shopping street.

The Cardo leads to the New Forum, a tremendous urban arena started in 203 CE by Septimius Severus. It is a monument to him, his family, and their pretensions, but although a work of colossal vanity expressing worldly ambition, it is also a heartbreakingly beautiful work of art.

The New Forum was – and remains – one of the greatest pieces of city landscape from the ancient world. The way of entry from the Cardo is through a small court, which suddenly opens in most theatrical manner into the main forum. Now the forum contains cascading mounds of dislocated classical architectural ornaments of the highest quality, but originally, on three sides, there were shady arcades formed by columns supported on arches, with the spandrels bearing huge and beautifully sculpted heads of the Gorgon. There was also a cult temple dedicated to Septimius Severus and the 'genius' of his family.

Opposite the temple is the basilica, which occupies one entire side of the forum. Basilicas were the most important public buildings of any Roman city, containing law courts, public meeting rooms, shrines, markets, and exchanges. This one appears to have incorporated a row of shops along its side facing onto the forum.

Basilicas – including the one at Leptis – are generally rect-angular in shape, having apses at one or both of the short ends, and double-height naves divided from narrow and often lower-ceilinged aisles by a screen of columns or piers. This basilica plan is now familiar because early Christians adopted it as the standard design of their churches, merely adding transepts to give the plan the shape of a cross.

The Septimius Severus Basilica – beautifully built of finely cut blocks of stone – is now a roofless and partly collapsed ruin, but it is still a magnificent building. The two tiers of columns separating nave from aisles now lie mostly in glorious fragments of heroic architecture, but the two apses on the short ends of the basilica retain much of their decoration, including pilasters bearing delicate carv-ings of Herakles and Dionysius, the patron gods of the Severan family.

The basilica was started in about 200 CE by Septimius Severus and completed over a decade later by his son, Emperor Caracalla, and so remains a personal testament to the power of the Severan family. But soon after 211 – when Septimius died in York while campaigning in Britain – this power started to wane and the family's story went horribly wrong. Caracalla proved a murderous ruler, executing his own brother and even attempting to kill him again after death by having him declared 'damned of memory'. In 217 CE Caracalla was assassinated in a most humiliating manner – while casually urinating by the roadside – and was succeeded as emperor by a Praetorian Guard named Macrinus. He lasted a year and was succeeded by Elagabalus, and then by Severus Alexander. Both were of the Severan dynasty, both were weak and useless as emperors, and both were assassinated. And so, in 235 CE, it all came to a sad and ignoble end for the Severan dynasty, with its tremendous but transient power now commemorated by these beautiful but tumbled stones.

73

PALMYRA

Palmyra, Syria: Colonnade Street | 1st century BCE – 3rd century CE

Palmyra was one of the great trading cities of the ancient world, a place of imperial ambition and – like Petra and Hatra – of spectacular architectural beauty. Then, just over 1,700 years ago, disaster struck and Palmyra, the 'Bride of the Desert', was laid to waste and gradually consigned to utter oblivion.

It was not until Rome conquered Syria in 64 BCE that Palmyra really started to flourish. Then the key trade route from the east moved north to Palmrya and Damascus, and in 217 CE Emperor Caracalla declared Palmyra a Roman colony – a most privileged status that relieved the city's merchants from the need to pay certain taxes. This produced wealth and soon vast and ornate buildings were completed, in a sophisticated classical style that made Palmyra one of the most beautiful cities in the empire.

As Palmyra increased in importance and blossomed in architectural beauty, the world around it descended into chaos, and old allegiances shifted. With the death of Emperor Severus Alexander in 235 CE, the Roman empire disintegrated into a quagmire of intrigue, infighting, and economic inflation that was to last for nearly sixty years – a time of turmoil, known as the Imperial Crisis, marked by the rapid rise and fall of over twenty short-lived emperors. In these circumstances, Palmyra had to fend for itself.

A member of a distinguished Palmyrene family, Septimius Odainat, rose to the challenge and supported Rome in its struggles, but he was murdered in 267 CE – along with his eldest son – in most mysterious circumstances. His wife, Zenobia, soon became the new power in the land, and to many it seemed that she had instigated the murder of her husband.

Rome was alarmed by this turn of events. The current Roman emperor, Gallienus, sent an army against Zenobia, but it was defeated. The stage was now set for an epic struggle. Zenobia was ambitious, ruthless, and had a lust for power, but she was also highly intelligent and had a deep love of the arts and culture. While she plotted her masterstroke – a move that would capitalize on the current weakness of Roman rule – she filled her court with artists, philosophers, and theologians. Rapidly, her plan revealed itself: she wanted not only to liberate Palmyra from Roman domination but also to establish her own empire in the Middle East, one that would challenge Rome itself, and to do this she aimed to grab parts of the disintegrating Roman Empire. First she seized the entire Roman province of Syria, and then, to the utter astonishment of all, in 269 CE she invaded Egypt and soon wrenched it from Roman control. Egypt had long been the 'breadbasket' of Rome, and the loss of such an essential asset was intolerable. But before Rome could react, Zenobia marched to grab more of Asia Minor. But a Roman army quickly retook Egypt, while Emperor Aurelian himself led the attack on Syria and in the summer of 272 CE Palmyra surrendered. Zenobia's fate remains debated, but her city survived, although reduced in status to no more than a garrison town following a subsequent rebellion in 273 CE

What survived above ground at Palmyra, when it was rediscovered by the west in the late 17th century, was just a fragment of the original city – the part that contained the public buildings and the richest and best stone-built structures. But even these fragments suggest most forcefully the beauty, scale, and architectural ambition of third-century Palmyra.

But these stunning structures, that impressed and inspired generations of travellers – both Christian and Muslim – were brutally damaged between May 2015 and March 2017 by the cultural terrorists of so-called Islamic State (IS) during its two brutal occupations of Palmyra.

Running roughly from west to east through the centre of the city is Colonnade Street, a processional route that must have been one of the most sublime streets in the ancient world. It had survived IS virtually intact. The street is formed by two parallel rows of giant Corinthian columns on which were placed statues of leading citizens, and behind which were shops and all manner of public buildings. The colonnades created a powerful architectural setting for life in the city, and today these rows of columns snaking through the desert are visually stunning and intensely evocative, speaking of lost urban grandeur and high ambition.

At each end of Colonnade Street are visually significant temples , and set along its route were two spectacular architectural features. At the east end is the mighty Sanctuary of Bel, within which was located the colonnaded Temple of Bel. Like virtually all the structures that survive, or survive until recently, the temple complex was built before Zenobia's time.

Built in around 30 CE, the temple, like its sanctuary buildings, offered a fascinating fusion of styles. It had a Greco-Roman-style Corinthian colonnade wrapped around its cella, but also details – stepped finials – inspired by earlier Assyrian architecture. The Sanctuary of Bel represented life, revealed by its location at the east end of the street, the place of sunrise and rebirth. The temple was severely damaged by IS, many of its stones obliterated. The Sanctuary wall and colonnades suvive.

At the west end of Colonnade Street is a pedimented funereal temple and beyond it the valley of tombs. So in one way Colonnade Street mediates between life and death; it is the passage from this world to the next. The funerary temple survives IS but – tragically – the best tomb towers have been destroyed.

Roughly in its centre of Colonnade Street was the Tetrapylon, a sort of omni-directional triumphal arch. Largely a late-20th-century reconstruction it had been laid low by IS. The Tetrapylon commanded vistas but also marked a turning point, for Colonnade Street, unlike comparable streets in Roman cities, is not ruthlessly straight but sinuous. Perhaps it speaks of feminine power, and of the great Palmyrene goddess Allat, who was the Arab equivalent of Venus. From the Tetrapylon the long vista of columns, with their relentless rhythm and grandeur, made a powerful impression - it still does for it offers glimpse of the experience of walking through this city 1,700 years ago. The experience would surely have been stupendous. At the west end of Colonnade Street was the late second century CE Monumental Arch, beautiful and delicate but laid low by IS, and to the south were the amphitheatre (slightly damaged), the agora and the senate, which survive unmolested. Sadly the same cannot be said of the jewel-like and almost perfectly preserved early second century CE Temple of Baal Shamin to the north of Colonnade Street. IS tumbled its stones with high explosives.

But Palmyra can yet be reborn, because in most cases its damaged monuments are capable of authentic reconstruction, All that is now required is the will and means to do so, and the opportunity.

74

PALENQUE
Mexico | c. 7th century CE

The roots of Maya culture stretch back nearly 3,000 years, but it coalesced into a coherent civilization with a consistent pattern of development from its 'classic' period, around 250 CE. It flourished, reached a peak of achievement in the seventh and eight centuries, and then – as was so often the way with the great civilizations of Central and South America – went into a sudden and mysterious decline and disappeared as a distinct entity. By 900 CE its great cities started to be abandoned, its people fragmented and absorbed – along with many of its beliefs and technologies – into subsequent cultures.

Palenque was built during the late 'classic' period of Maya culture, during the hundred years after 650 CE, by King Janaab' Pakal and his sons. It grew to great size, incorporated architecture of highly sophisticated design and engineered construction, and then, between 850 and 900 CE, was abandoned.

Pakal's tomb lies in the heart of Palenque's greatest structure, a burial pyramid now called the Temple of the Inscriptions. When this tomb was discovered in 1952 it was soon realized that this was the most important pyramid burial yet found in the Americas.

The first thing that strikes the visitor is the scale and design of the pyramid and its relation to the landscape and to the city. Palenque stands on high land that commands a fine view over the plains below, so the occupants of the city were well placed strategically to detect attack and control movement over their land. The site of the city, although to some extent a plateau, is also undulating, with some of the structures placed on natural mounds or, as in the case of the Temple of the Inscriptions, to the side of rising ground.

Unlike many Central and South American civilizations, the Maya evolved writing – glyphs with a phonetic value – that expressed their rituals, traditions, and history. From this writing and from other sources, much has been learned about Pakal and the Maya of Palenque. For example, it's known that Pakal died in 683 CE and before his death organized the creation of his sarcophagus, tomb, and burial pyramid. Also from deciphering Maya texts it is known that a funerary pyramid was called a 'burial hill', a term that confirms what appears to have been a universal notion in ancient religions, that high lands were the domains of the gods and that pyramids were man-made sacred mountains.

The pyramid has nine levels or steps, no doubt reflecting the Maya belief that there are nine levels to the underworld. The Maya perceived creation as a trilogy and, strange as it may seem, this creation was symbolized by a cross, representing the World Tree. Each level of the underworld, presided over by a different god, was represented by the lower vertical portion of the cross. The land of the living was represented by the horizontal portion of the cross, and finally, the thirteen levels of heaven were represented by the upper vertical portion of the cross.

Cutting through – and up – these nine levels of Pakal's step pyramid is a steep staircase that, presumably, represents the journey of the deceased's spirit to a temple on the summit of the structure. From here a second staircase plunges down, within the body of the pyramid, to the burial chamber and sarcophagus.

Constructing this staircase passage within the mass of the stone-built pyramid was an extraordinary feat of engineering. The roof of the passage, as the stairs descend, has to support the increasingly heavy load of the masonry of the pyramid above. To achieve this, the Maya evolved the corbel vault. This is a splendid and sophisticated construction in which horizontal stone slabs project from a wall, one upon the other, to create a massively strong V-shaped roof or vault. In the case of this pyramid, the vault transfers the massive loads above the staircase passage to the thick walls on each side of it.

Roughly halfway down there is a landing where the staircase turns 180 degrees before continuing downwards. It's hard to be sure, but it's possible that the staircase was constructed in thirteen stages, each stage defined by a change in level of the surface of the vault, so that the descent represents a journey through the Maya heavens.

At the bottom of the staircase is massive triangular stone slab – the door that sealed the king's tomb. Beyond is a vaulted tomb chamber containing Pakal's sarcophagus, which is so large that it is evident the pyramid was built around it. The tomb is rich with imagery, revealing Pakal's power and hopes and much about Maya theology, and in the centre of the stone sarcophagus is cut a strange shape. It is to receive the body of the king, but the recess is not body-shaped. It has been called the shape of a uterus – a womb.[1] So in death Pakal believed there was new life. For the Maya there was no such thing as death, only transformation. Seemingly Pakal had his body placed in a womb of stone as a sign of the certainty of his rebirth, and the cover of the sarcophagus shows the World Tree in the form of a cross, sprouting from his body as corn sprouts from a seemingly dead and dry husk. A glyph of the sun is also depicted on the cover, suggesting that, like the Sun, Pakal would rise again after his journey through the 'otherworld'.

So Pakal assumed that, in harmony with the life and death cycles of nature, he would rise from his tomb and ascend the inner staircase. Certainly the immense structural strength of the staircase suggests that it was built to last – to allow Pakal a chance of rebirth, no matter how long it might take.

75

MACHU PICCHU
Peru | c. 1450–75

Machu Picchu, high in the Andes, is the ideal 'lost city', calculated to inflame imaginations and fuel fantasies. Its location is spellbindingly beautiful; it was created by a people that suffered a tragic and rapid eclipse; and its architecture and planning express a way of life that seems parallel to, yet distinctly different from, the familiar thrust of history that is deemed to have determined mainstream cultural development – notably, it was an advanced society that did not make use of the wheel or develop a written language, and that had no iron-based technologies or artefacts. Lastly, the remote city, abandoned centuries ago in mysterious circumstances and in haste, was not rediscovered by the outside world until the early twentieth century.

What was immediately – startlingly – clear when the city was virtually stumbled upon in 1911 by an American academic named Hiram Bingham was that its stone buildings were remarkably well constructed, that the city was harmoniously planned, and that its remote site, perched high in the mountains, offered not only defence but also closeness to the gods. It was evidently, among other things, a sacred city.

Machu Picchu retains a sense of mystery. It was forgotten by the outer world for nearly 400 years, and even now its primary purpose is not certain, nor is it known for certain why and when it was abandoned. What is agreed is that Machu Picchu was constructed in the mid- to late-fifteenth century by the Incas, during the golden age of their empire, and that it was not discovered by the Spaniards after they had crushed the Inca armies in the 1530s.

The first view of the ruined city, if the Inca path is followed, is one of the great sights of the world. Suddenly a corner is turned and there are the ruins, nestling on a natural shelf – extended by Inca engineering – between a series of peaks. One of the most satisfying things, looking down on the ruins, is that they represent a clear diagram of Inca urban ideals. The highest point within Machu Picchu is very clearly occupied by a temple, the altar of which is visible and carved from the peak of the mountain on which the temple stands. From the little that is known, the Inca regarded Machu Picchu as a cross between a royal city and, in Western terms, a monastery – a preserve of the high caste and the holy. But it was also a functioning city, for it is still surrounded by steep terraces on which the working population grew a large proportion of the crops needed to make Machu Picchu a self-sustaining community.

The built-up area is divided into two parts by a long and wide-terraced esplanade. Both halves contain residential zones, comprising short streets running along stepped terraces. The lower terraces are lined with small individual houses, with continuous rows of houses along the upper terraces – literally terrace houses. Of these two residential zones, the higher was the grander because it incorporates the royal palace and the temples. In addition there were storehouses, bathhouses, and an impressive water system.

The surviving fabric of Machu Picchu reveals a most curious and individual architectural tradition. Wall construction is superb, ranging from neatly coursed and roughly squared stone blocks for modest building to construction often incorporating vast blocks of stone that are flat-faced but multiplaned rather than squared, with blocks locking together like giant jigsaw pieces. This complex construction was probably a response to the threat of earthquake. But the Incas did not develop the arch or the dome, which means that even the main gate to Machu Picchu is provided with only a horizontal stone lintel, making its width very narrow – more like the door to a room than to a city.

The architectural glories of Machu Picchu are its temples. Nestling among the high-class housing, near the small royal palace, is the Temple of the Sun. This is a truly remarkable structure with a tapering window that allows a ray of light to enter at dawn on the winter solstice and fall across the altar, again cut from the rock on which the temple stands. The temple's walls, in this case made of rectangular blocks with perfectly straight edges, wrap around the altar stone, implying that it's the heart of a cosmic spiral.

The highest temple in Machu Picchu, perched on a mountain peak, is the most moving of all. At its threshold is a boulder carefully carved to mimic, in miniature, the profile of the group of mountains beyond, and in the centre of the now roofless temple is a free-standing altar carved out of the actual mountain peak. Together these offer compelling evidence of Inca veneration of nature, and suggest that this altar, carved from a sacred mountain, was the Axis Mundi – the axis around which the world turned and the very centre of the Inca universe.

76

PLACE DES VOSGES
Marais, Paris, France | 1605–1612 | Perhaps Jacques Androuet du Cerceau II

The Place des Vosges offers a most satisfying urban experience. Its magic comes from a combination of things, ranging from the general to the very particular. The scale, ambition, and nobility of the conception are deeply impressive – the creation of an ideal city square or 'place' to house the grandest in the land within what appears to be an evocation of the inner courtyard of the legendary Temple of Solomon. It was an audacious enterprise, but this vision was to be realized with a certain humility. Although a place of individual palaces for royalty and courtiers, more important was the unity of architectural expression achieved through coherence of design. This meant that, by and large, the design of individual houses was subservient to the creation of a uniform composition with the overall character of a single urban palace. Only the pavilions in the centres of the north and south ranges, intended for the king and queen of France, proclaim their individuality, although within the constraints of the architectural language of the scheme.

Combined with this palatial imagery and Solomonic association was a sense of the antique, for the individual buildings incorporate a continuous arched and vaulted loggia. Such an arrangement was not novel: arcades had been included in medieval urban architecture, were pioneered in Renaissance style in Florence in the early fifteenth century, and were used to great and influential effect in 1594 by Bernardo Buontalenti in the Piazza Grande, Livorno, Italy. At the Place des Vosges, as perhaps in Florence and Livorno, the arcade was intended to give the design the ambience of a Roman forum, which was almost invariably framed by colonnades providing covered walks.

The vaulted arcade at the Place des Vosges is one of its most successful features. Its generous width engenders a most civilized form of urban life, offering a place of pleasant promenade or repose, and the relative lowness of its shallow vaulted ceilings, the stoutness of its piers, and the glimpses offered through its arched openings give it the feel of a shady, sheltered, and almost sacred grotto within the hustle and bustle of the larger city.

The intimate shelter of the arcades within this public parade of palaces is just one of the Place's intriguing contrasts. Another is the material of construction, which – as was the fashion in early seventeenth-century northern France – combines homely red brick with crisply cut white stone. Also there is now the contrast between the lofty and intensely urban elevations and the verdant gardens they embrace. This adds much to the magic of the Place, although a later development.

The Place was commissioned by King Henri IV in 1605 and was the key move in his attempt, as the first Bourbon, to give Paris the hallmark of a modern royal capital. Shortly before his assassination in 1610, Henri commissioned the Place Dauphine on the Île de la Cité, which was designed in the same uniform manner as the Place des Vosges.

The identity of the architect for these two great projects is still debated. Possibly it was Jacques du Cerceau II, who since 1596 had been the king's comptroller of building. The plan of the Place is curious – it's a perfect square with sides 140 metres (460 ft) long, or in ancient measure, just over 300 cubits. The courts and halls of Solomon's Temple were square or cubical, or simple permutations of these proportions, and according to the Book of Genesis, Noah's Ark was 300 cubits in length (Genesis 6: 15), so Biblical precedent for the design of the Place seems likely.

While the antecedents of the Place are fascinating, its influence was far-reaching and profound. It was arguably the first great residential urban square of uniform classical design, with the Plaza Mayor in Madrid, started in 1617 by Juan Gómez de Mora, perhaps being its immediate progeny. However, the origin of this great Spanish square is enigmatic, for such a creation had been conceived as early as 1577 for Philip II by Juan de Herrara, who in 1584 had completed the Escorial.

The legacy of the Place des Vosges in Britain is more certain and more direct. In 1625 Henri IV's daughter, Henrietta Maria, married King Charles I, and this strong-willed woman, with a penchant for architecture, must have been influential in the Earl of Bedford's creation in the early 1630s of the Piazza in Covent Garden, London. Designed by Inigo Jones, this development, which required a licence from the king, incorporated uniform red-brick elevations laced with pale stucco ornament and ground-floor arcades that were clearly inspired by the Place des Vosges and the piazza in Livorno. Henrietta Maria was also involved in the promotion of other visionary buildings in London, particularly with her close adviser the Earl of St Alban, who developed the aristocratic and uniform St James Square in 1662, on crown land – a project for which he would later be dubbed the 'Father of the West End'.

These prototypical London developments became, themselves, models for much admired Georgian urban creations in Britain and its colonies, confirming the profound influence of the Place des Vosges on world architecture.

77

THE CIRCUS
Bath, England | 1754–68 | John Wood the Elder

Bath is one of the great classical cities of the world. Of Roman origin, Bath in the Middle Ages was a monastic and market town with an economy based on the wool industry, but it flowered in spectacular manner during the eighteenth and early nineteenth centuries to become England's premier city of pleasure and elegance. Bath's architecture was conceived, in almost theatrical manner, as a backdrop for its revels, yet came to represent an ideal of urban design and living.

Much of the impressive architectural character and apparent visual unity that Bath possesses was achieved through a series of happy accidents. In the 1720s a small number of visionary, powerful, and influential men – notably the architect John Wood, the land and quarry owner Ralph Allen, and the landowner Robert Gay – collaborated to ensure that local limestone became the preferred building material and that the then nationally fashionable Palladian style of classical architecture became the house style of the rapidly expanding city.

The uniting of individual buildings within a uniform and palatial urban elevation has European precedent – for example, the street-like courtyard elevations of the Uffizi in Florence of 1580, by Vasari, and the Places des Vosges, Paris, of 1605–12 – and was pioneered in England in the 1630s with Inigo Jones' Piazza in Covent Garden, London.

Jones' development – produced for the Earl of Bedford – was largely residential but contained enough other uses to give it the feel of a small, self-contained town. It also reflected a hierarchy of occupation. Large houses of uniform external design framed the Piazza, which contained a significant public building – a church – much like the arcaded Piazza Grande of 1594 in Livorno, Italy. Smaller, less architecturally formal buildings with a mix of uses were located in subservient streets forming a small orthogonal grid around the square.

The Covent Garden model was taken up and developed by other London estates in the early decades of the eighteenth century – notably Cavendish Square from 1717 on the Cavendish Harley estate and Grosvenor Square from 1721 on Sir Richard Grosvenor's Mayfair estate. This architectural and planning approach to urban development was continued – indeed, in a sense given its finest and most complete expression – with Queen Square in Bath, designed by, and developed from 1729 to 1736 under the control of, John Wood the Elder.

The cultural references applied by Wood in the design of Queen Square are rich and complex. These not only include the precedent of the Roman forum and Biblical descriptions of Solomon's Temple but also reflect contemporary perceptions about the form and dimensions of such inspirational ancient British monuments as Stonehenge and the stone circle at Stanton Drew, near Bath. The notable feature of the square is its north side, which was conceived as an architecturally uniform group of individual houses designed to look like a single palace elevation, with central pediment and end pavilions, realized in beautiful local limestone from quarries owned by Ralph Allen.

Also important was the method by which Queen Square was created. Wood entered into a development agreement with the landowner Robert Gay, drew up designs that determined the size, the elevations, and the quality of the construction of the houses, and then sub-let plots to speculating builders for a specified number of years.

Queen Square was a success and did much to establish Palladian design as the preferred Bath style. One of the reasons for this success was that Wood's architecture was imbued with patriotic fervour, centred on a declared desire to 'revive' an ancient British culture inspired by Stonehenge and

Stanton Drew, whose measurements he had applied in the design of Queen Square.

Wood also applied these measurements, combined with the circular forms of these ancient monuments, when designing the Circus in 1754, for land on the Walcot Estate that had been owned by Robert Gay and, after his death in 1737, by his daughter, Margaret Garrard.

Wood, along with many of his contemporaries, viewed the ancient and mythic King Bladud of Bath as a real being who had brought Greek classical culture to Britain long before the arrival of the Romans. The antiquary William Stukeley had been a pioneer in the quest for a distinct and proud national identity rooted in the remote past. He was involved in reviving interest in druids and other aspects of distant and mythic British history that suggested ancient, perhaps classical, wisdom.

To a significant degree Wood's buildings in Bath were part of the movement to discover and assert an ancient British identity, and so can be viewed as architecture bathed in heroic meaning. In his *Essay Towards a Description of the City of Bath* of 1742, Wood wrote of finding what he took to be ancient altar stones in a field near Lansdown, Bath; he believed this was evidence that it had once been the site of a temple, like Stonehenge, which he argued had been dedicated to the sun god Apollo. As Wood put it, 'as we ascend the Hill now bearing the name of Lansdown, there are three large stones lying upon the Ground, in a little Field by the side of the road, known by the name of Sols rocks, with a Foundation just behind them, shaped into a Circular Form … These three Stones, when erect and perfect, seem to have made a stupendous Altar; and the circular Foundation behind them seems to have borne other erect stones, which, in all Probability, were set up by King Bladud for a Temple in honour of the Sun'.[2]

This discovery seems to have inspired Wood to build his own round temple on the site – the Circus – and then to propose a crescent-shaped temple to the moon on an adjoining site. After Wood's death in May 1754 (just as construction of the Circus was about to start), this crescent-shaped development was eventually realized, between 1767 and 1774, by his son, John Wood the Younger, and called the Royal Crescent. For Wood the Elder at least, the connection was clear between King Bladud, Minerva, and Celtic myth and religion, and the creation of the Circus and the Royal Crescent.

The Circus was conceived as three curved segments set on high land north of the city and so enjoying fine and fresh prospects. The geometry is fascinating, and not only suggests connections to Stonehenge and Stanton Drew (key dimensions of the Circus echo dimensions at Stanton Drew) but also appears to relate to the rituals and imagery of Freemasonry, a highly respected institution in eighteenth-century Britain.

The three points of entry into the circle of the Circus define a triangle, and for Masons a triangle set within a circle was a most important symbol, representing the Christian Trinity set within the circle of eternity. Indeed, the very form of the Circus – and its relation to Queen Square – can, among its many possible attributes, be seen as Masonic. From the 1720s in Britain, most urban planning was characterized by the use of right-angular grids of streets incorporating squares, such as London's Grosvenor Estate. So Wood's introduction of a circus was novel enough, but in the context of Queen Square it is positively bizarre. Seen in plan, with the Square connected to the Circus by a street extending not from its centre but from one side, the composition forms the outline of a key, which was an important Masonic symbol representing access to arcane knowledge.

The design of the elevations of the houses is extraordinary and packed with potential meaning. They are uniform and ornamented with three tiers of orders, each separated by entablatures with Doric at ground level, then Ionic, and finally Corinthian. The common contemporary opinion was, no doubt, that expressed in 1771 by Tobias Smollett through one of his characters in *The Expedition of Humphrey Clinker*: 'The Circus is a pretty bauble, contrived for shew, and looks like Vespasian's Amphitheatre turned inside out.'[3] Its tiers of classical orders of course suggested, to most educated viewers, the inspiration of the Colosseum – but what could they have made of other details, especially the strange emblems in the Doric frieze at ground-floor level? Recent research has shown that the majority of these were taken from George Wither's *Emblem Book* of 1635.[4] Why Wood used this source is unknown, but presumably it had a meaning beyond the mere weird and decorative. Certainly, a number of the emblems – notably a key and a beehive – can be seen as Masonic, so it is likely that Wither's book was embraced by eighteenth-century Masons as a seminal text. More obvious but equally curious are the finials that sit on the parapet of the Circus. By tradition these should be pineapples – a symbol of welcome – but here they are acorns. These were a prime druidic symbol in the eighteenth century because druids were known as the 'men of oak', for whom the oak tree was sacred.

Far from trying to use the stupendous structure of the Colosseum as an unlikely model for speculative houses,

Wood was, with the designs of Queen Square, the Circus, and the Royal Crescent, attempting to recreate the planning and architecture of ancient Britain.

78

SYDNEY OPERA HOUSE
Sydney, Australia | 1957–73 | Jørn Utzon; Hall, Todd & Littlemore

The Sydney Opera House is one of the most famous, most artistically striking, and most notorious buildings constructed in the twentieth century. The intention behind its creation was to help put Sydney – indeed, Australia – on the cultural map. The acquisition of an opera house has long been recognized as an expression of a city having attained cultural maturity – for example, the opera house at Manaus, constructed in the 1880s on the River Amazon and deep in the rainforest of Brazil by ruthless and rich rubber barons seeking cultural respectability.

In its aims and in its architecture the Sydney Opera House has proved a success way beyond the dreams of its creators, and it remains not only the best-known modern building in Australia – and a token of national pride and identity – but also one of the best-known twentieth-century buildings in the world. Its familiarity and success are due largely to its striking exterior sculptural form that works brilliantly with the sparkling light of its waterside setting.

But this great success has been achieved despite the bitter struggle surrounding the design and construction of the building, which included its Danish architect Jørn Utzon walking off site during construction and leaving the country. Utzon was overwhelmed by the mounting complexity and cost of the project and felt undermined by what he perceived to be a lack of understanding and support from the political authorities responsible for realizing his inspiring and visionary design. Never, surely, can a much-loved national icon and generally recognized architectural masterpiece have been forged in such unpromising circumstances.

The story of the conception and construction of the Sydney Opera House is one of the world's most gripping architectural sagas, offering insights into the complex problems of creating a great public building burdened with the responsibility of expressing the honour and dreams of a nation.

In the time-honoured tradition of public patronage for the creation of a building of intense national importance, the design was sought via a high-profile competition that Jørn Utzon won in January 1957. The cost estimate for the building at the time was A$7 million, and the engineer chosen to help Utzon execute his revolutionary design – notable for its free-flowing organic form, with vast and structurally daring steel reinforced concrete shells – was the Anglo-Danish and London-based Ove Arup. Utzon's architectural inspirations were the Scandinavian architects Erik Gunnar Asplund and Alvar Aalto, who, Utzon observed, worked not in two-dimensional plan but in three-dimensional space. The building's design was a true child of Aalto, conceived in flowing volume.

Much design work and many calculations were needed to refine the winning design and make it buildable, but there was a sense of urgency among politicians, and the New South Wales premier, John Cahill, insisted that construction work start by February 1959.

By September 1961, before the complex construction of the concrete shell roofs had started, the estimate for the completion of the building had risen to A$18.6 million. At this point Utzon and Arup were still pondering the geometry of the shells, and in October 1961 Utzon decided that – for aesthetic and structural reasons – they should be based on the form of a sphere. So the more organic form of the original design gave way to a series of co-axial parabolas, a form that the engineer could define mathematically, allowing him to calculate the forces at work and suggest a structural strategy for construction.

By April 1962 the estimated cost had risen to A$27.5 million. In March 1963 construction of the shells started, and by July 1965 the estimated cost of completing the Opera House had risen to a staggering A$49.4 million. At this stage, six years into the project, with much still to do and over budget, an air of gloom set in; by early 1966, this had turned to panic and recriminations. There was talk of withholding Utzon's fees, of handing the project over to a cost-cutting drafting office, and of exerting more political control over finances. In February 1966 Utzon threatened to resign, a gesture that was immediately accepted. He flew out of Sydney in April, leaving confusion behind him. Architects Hall, Todd & Littlemore were appointed to take over, with Peter Hall largely responsible for interior design, and they quickly concluded that Utzon's proposed dual-purpose hall was unworkable. Flexibility was reduced as the potential for staging opera was dropped from one of the halls.

The Opera House was pushed forward at greater speed and eventually opened by Queen Elizabeth II in October 1973 – nearly fifteen years after construction had started, six years behind schedule, and at a cost of A$102 million, more than ten times the original estimate.

The execution of the Sydney Opera House and the integrity of its design were compromised by cost control, political wobbles, and Utzon's absence – as had been predicted in March 1966 by the vociferous 'Bring Back Utzon' campaigners. A particular loss are elements of the flowing, organic interiors envisaged by Utzon, which has led in parts of the building to a strange sense of disassociation between startling exterior and slightly less startling interior.

But from what could have been a tragedy emerged a triumph – and one marked eventually by reconciliation and recognition for Utzon's remarkable achievement. As with late Le Corbusier designs such as the chapel at Ronchamp, Utzon's creation re-imagined – in a sense, reinvented – twentieth-century Modernism, helping it to evolve from box-like primary functionalism to something more humane, emotive, sinuous, and artistically rich. In 1978 Utzon – aged 85 – was awarded the prestigious Gold Medal by the Royal Institute of British Architects. The citation confirms that the Sydney Opera House possesses 'great beauty' and is a compelling 'symbol for not only a city, but a whole country and continent'.

CHAPTER 5

BIG AND BEAUTIFUL

I N ARCHITECTURE, SIZE CAN matter. This is especially the
case if the building is designed to express wealth, power,
identity, or military prowess, or if it is a sacred structure
dedicated to the assertion of divine verities, to proclaimthe
glories of the dead or the eternal qualities of a deity or religious belief. The Great Wall of China is the prime example
of vast scale in the service military objectives; the Forbidden
City in Beijing suggests the connection between largeness
and heaven-ordained and guided rule, and the Great Pyramid
at Giza, Egypt, confirms the importance of scale in the
creation of a building that enshrines some great mystic or
portentous secret, perhaps touching on the nature of life,
death, and afterlife – on the past, the present, and the future.

In recent centuries, size has acquired an additional value,
to do with mammon – with material wealth rather than god.
Vast volume can now be a measure of a building's importance when a structure's value is assessed not by its artistic or

engineering achievements, by its beauty, sacred intent, ingenuity or meaning, but through its commercial real-estate
potential and capacity to contain a maximum amount of
lettable space. The highest structures in medieval Europe
were cathedral towers – for example that of Ulm Minister in
Germany, constructed from 1377 to 1890, rises 161.5 metres
(530 ft) making it the tallest church spire in the world. In the
Middle East and Central Asia, the pinnacles of man's architectural aspirations were minarets.

But, by tradition, size of buildings – particularly height
– has been viewed with some suspicion. If done merely for
worldly glory, such hubristic efforts to outdo the wonders of
God's creation can lead only to fatal nemesis. The Biblical
Book of Genesis includes the story of the construction of the
Tower of Babel and describes how men gathered together
and said, 'let us build ourselves ... a tower whose top is in
heavens; let us make a name for ourselves.' But the Lord,
when He 'came down' to see the tower 'which the sons of
men had built', was not impressed. Indeed He seems to have
been alarmed and supposed that 'now nothing that they
propose to do will be withheld from them'. So, to prevent this
eventuality, the Lord 'scattered them' and confused their
language so that they could not understand each other and so
could not combine to complete the tower or attempt such
another outlandish monument to overweening pride.

Biblical lessons have not of course been learned. Towers
rear ever taller and taller and now the most aspiring are not
gracing cathedrals or mosques but are commercial structures
containing offices, hotels, retail uses, restaurants, or ultra-ex-
pensive apartments. The tallest structure in the world is
currently the Burj Khalifa in Dubai, which contains a mix of

residential and commercial space and was built between 2004 and 2009. It rises 830 metres (2,720 ft) and was designed for Dubai and United Arab Emirates clients by Chicago architects SOM with Adrian Smith and engineer Bill Baker.

79

GREAT PYRAMID AT GIZA
Cairo, Egypt | Perhaps 2550 BCE

The Great Pyramid at Giza is now associated with the Old Kingdom pharaoh Khufu (or Cheops), who is thought to have died around 4,580 years ago. It was perhaps constructed about a hundred years after the Step Pyramid at Saqqara, and is far more precise in its setting-out and construction, and vastly more complex in its form and probably its meaning. It is also far bigger. Indeed, the Pyramid of Khufu is one of the largest man-made structures of the ancient world, which, combined with its precision of construction, gives it a truly miraculous quality.

This sublime scale and perfection of form and construction have long been acknowledged. As far back as the second century BCE the Great Pyramid was regarded as both venerable and superlative when it was listed as one of the Seven Wonders of the World. It is the only survivor, in situ and largely intact, of these ancient wonders

The Great Pyramid stands with two other pyramids – slightly smaller and believed to be later – on a plateau raised slightly above the city of Cairo, which now encroaches in an ugly and disrespectful manner. Even without the contrast of the chaotic and rambling modern city, however, it is evident that these three vast and elemental structures embody geometrical perfection, beauty, and harmony.

It is also clear that all three pyramids – Khufu, Khafre, and Menkaure – are precisely aligned, and this alignment relates directly to the form of the Great Pyramid. Its sides rise at an angle of 51.52 degrees, but they are now ragged, formed with blocks of limestone, having been stripped of their smooth facing stones more than 700 years ago. It must be assumed that the original angle was between 51 and 52 degrees, which is fascinating because 51 degrees is also the angle defining, with impressive if not absolute precision, the relationship between the southeast corners of the three pyramids. So it would seem the pyramids are geometrically related and, evidently, for their builders the angle of 51 degrees had special significance. Exactly what this significance was remains one of the many mysteries of the pyramids.

Others mysteries are more basic and mind-boggling, making the Great Pyramid, in particular, the perpetual victim of inflamed imaginations. Like the neighbouring Sphinx, it embodies a mighty riddle. Although the past 200 years or so have seen the fabrication of an 'orthodox' chronology and purpose for the Great Pyramid, according to which it was constructed around 4,550 years ago for Khufu as his tomb and perhaps sun temple, the fact is that this is nothing more than a consensus based on informed speculation and interpretation. Even a quick glance at the Great Pyramid's vital statistics and main physical characteristics raise some basic questions about this hypothesis. The obvious point – often taken for granted – is the structure's astonishing scale. In this respect it was unprecedented; it was like a force of nature created in defiance of time and intended to outlive humanity.

And then there is the phenomenal accuracy of its setting-out. The four corners of the Great Pyramid define the four cardinal points of the compass with an accuracy of within one twelfth of a degree, a remarkable precision that reveals the use of a lost technology or a great understanding of the heavens. This celestial wisdom is supported by many other aspects of the Great Pyramid – for example, it has been calculated that the main, and originally hidden, entry to its interior passages is aligned on the position that the Pole Star would have occupied around 4,500 years ago, while other details align with stars in the constellation of Orion.

The Great Pyramid rises 138.8 metres (455.4 ft), but it was 146.5 metres (480.5 ft) high before it capstone – the pyramidion – was removed. This stone was a little pyramid in itself, the symbol of the Benben (the mound of primordial creation), and had a sacred purpose. If, among other things, the pyramid was a temple to the sun god Ra, then the pyramidion marked the dawn of each day, for it caught and, with its skin of gold (or of a gold and silver alloy called electrum), reflected the first rays of the dawning sun, heralding Ra's daily rebirth.

The pyramid covers 5.2 hectares (13 acres) of ground and contains around 2.3 million blocks of stone, each weighing on average 2.5 tonnes, rising in 203 courses. The total weight of the stone used was about 7.5 million tonnes. The stone was of three types: granite transported about 1,000 kilometres (600 miles) along the Nile from Aswan in the south, limestone from Giza, and harder limestone from across the Nile at Tura for the polished stone skin that originally encased the pyramid but has now been almost entirely looted.

The accuracy of the length of the sides is most astonishing, ranging originally from an estimated 230.25 metres to 230.39 metres (755.4–755.9 ft), and the base or horizontal height level of the pyramid differs from corner to corner by only 2.1 centimetres (0.8 in). This is accuracy beyond comprehension – not only more than would be deemed necessary in a comparable modern construction, but probably more than could now be achieved.

The real question, however, is not how such painstaking accuracy and perfection was achieved, but why. The structure does not require it for practical or even for visual reasons. Only a machine with a very precise and demanding function could require such accuracy of fabrication and fine-tuning. But if the Great Pyramid is a high-performance machine, what on earth was – or is – its function?

And this dazzling accuracy at huge scale was achieved – if the orthodox view is accepted – using conventional, and fairly primitive, Bronze Age building technology. This means that the hard blocks of granite, with their long and perfectly straight edges and immaculately flat and smooth faces, had to be cut or pounded into shape using blocks of diorite as 'pounding stones', or soft bronze chisels – it's theoretically possible, but it's hard to comprehend how the existing accuracy could have been achieved this way. Bronze chisels would have been required in their tens of thousands, and relatively few have been found in Egypt. Even the softer limestone would have been extremely difficult to shape in bulk and accuracy with such limited technology. The very few surviving facing limestone blocks have straight edges and incredibly tight joints less than 0.5 millimetres thick. As the great archaeologist William Flinders Petrie observed after his logical and expert analysis of the Giza pyramids in 1880, the accuracy of fabrication and placing of the facing or 'casing' stones was 'equal to opticians' work, but on a scale of acres'.

The construction project is truly vast, outstripping virtually any other comparable single-building scheme in the ancient or modern world. Was it really all just to bury a pharaoh and, perhaps, to guide his soul to the next world? It is possible. Pharaohs regarded themselves as the children of the gods – indeed, gods themselves – and they were accepted as such by their subjects. To honour a dead pharaoh was thus to honour the gods, in which case the pyramid would be simultaneously a great tomb and a great temple.

The complex and seemingly contradictory passageways and chambers inside the Great Pyramid have never been explained. Theories are many, but none are utterly convincing. The system appears to be loaded with symbolism, and it's easy to see it as representing a journey of the soul – even as a 'machine' for rebirth – with routes deviating from and off the true ascending passage, as if representing the temptations and hindrances that can dog the lives and aspirations of human beings. But what is certain is that the system is magnificently constructed using huge granite blocks that, in their weight, precise form, and razor-sharp joints, are utterly astonishing. How were they cut and lifted into place with such unerring accuracy? We really have no idea. And how did the builders learn to give strong materials greater strength through skilled design? Their engineering brilliance is astounding. They knew how to construct this system – even the long shafts that dart in mysterious manner north and south from inner spaces to offer glimpses of the sky and stars – to withstand the massive compressive weight of the pyramid's masonry body. The Great Gallery, with its ceiling supported on corbelled walls, is one of the greatest pieces of engineered architecture of the ancient world.

Beyond the Great Gallery is a chamber that, it's assumed, was for burial. The chamber is an almost perfect double cube in volume. This is one of the key proportions of sacred architecture and of tombs found around the world, and is one of the seminal proportions ascribed in the Old Testament to Solomon's Temple. Perhaps it all started here. And the granite chest in the chamber – so large that the pyramid had to be built around it, and assumed to have been Khufu's sarcophagus – was a container not for a body but for sacred objects, comparable to the Biblical Ark of the Covenant that the Jews considered the repository of immense power and the secrets of creation. Interestingly, the sarcophagus is again roughly a double cube in proportion, much like the Biblical description of the Ark, which was also as high as it was deep.

The Great Pyramid is not only the oldest nearly complete building on Earth but still, in many ways, the greatest, and the one most able to engage and excite the imagination.

80

GREAT WALL OF CHINA
220 BCE to 16th century

The Great Wall of China is almost more an idea than an actual object. Famed as the largest man-made construction on earth, it lives in the imagination – an emblem of what man can achieve with ruthless organization and commitment. In fact the Great Wall is not a single object but many. It's the collective name for a series of defensive walls – made of stone, rammed earth, wood, and brick – dating back to the seventh century BCE and all designed to protect the borders of a nation from marauding nomads bent on mischief. So long is this collaborative creation that it has proved almost impossible to measure, although current estimate is that the 'wall' measures around 21,196 kilometres (13,171 miles) in length.

The most famous section of the wall is that built – and rebuilt – between 247 and 210 BCE by Qin Shi Huang, the First Emperor of China, who unified and expanded earlier frontier walls into a national defensive work, but virtually nothing of the First Emperor's wall has survived intact. The best-known surviving portions of the Great Wall are those built near Beijing during the Ming dynasty, which lasted from 1368 to 1644. A section of this Ming wall at Jinshanling, about 125 kilometres (78 miles) northeast of Beijing, is particularly well preserved – it snakes in an exciting manner over mountainous terrain and appears to be in a remarkably intact state. In fact it has been much rebuilt in recent years, but generally in an archaeologically correct and authentic manner.

The Great Wall is an object of immense pride in China – it is a massive and early technical achievement, and it defines China's ancient identity as a nation and culture. It expresses China's historic continuity and is taken as physical proof that the civilization of modern China stretches back at least 2,500 years to the age of Confucius. Although the First Emperor did not initiate the construction of the Great Wall, he did, by making it into a coherent defensive system, give security and identity to the various states he had amalgamated into an empire that became the core of modern China.

It is said to have taken ten years and 30,000 men to build the wall constructed during the First Emperor's time, with the bodies of thousands – dead of exhaustion, or sacrifices – mixed into the brick clay and mortar from which much of the wall was built. For this reason, the wall – arguably the longest graveyard in the world – is still known by many Chinese people as the 'Wall of Tears'.

The 10.5-kilometre (6.5-mile) section of wall at Jinshanling was constructed – or reconstructed – in about 1570. Approaching the wall from the outside, one sees it from the perspective of the nomadic raiders that it was designed to repel.

The engineers who built this section of the Great Wall used the land well. It twists and turns to occupy ridges, to make the most of high points and natural gullies to command the land before it. In fact the wall was only the main section of a system of defence in depth, for, at regular distances, isolated towers are set strategically on high points in advance of the wall. These were to operate as strong vantage points, with their garrisons offering an early warning of attack to the soldiers on the wall. The wall itself is beautifully made. It rises about 8 metres (26 ft) high and here is made of large, square-shaped, hard, kiln-fired pale-brown bricks that rise off foundations laid on the solid rock or, occasionally, off lower courses of masonry.

The defensive system is fascinating. The wall-walk is about 6 metres (20 ft) wide – like a modest street – with the wall at its base substantially thicker. Along its length the wall-walk has parapets on both sides. Clearly the designer of the wall envisioned a situation where a section of the wall was breached and the garrison on the wall could conceivably come under attack from both sides simultaneously. This eventuality is also responded to by the design and location of the towers placed regularly and at relatively short distances along the wall. Each of these towers would have contained its own small garrison, and each is a small, self-contained castle. So, if one section of the wall fell to attackers it could, in theory, be isolated as the defenders retreated into the towers on each side. This makes the idea of a long defensive wall more reasonable – if one part falls, not all is lost, for the wall is in fact made up of a large number of virtually independent mini-fortresses.

The wall, the main spine of a system of in-depth defence, was a perfect military machine well calculated for the enemies it faced. These were nomads who fought from horseback, used shock tactics, favoured hit-and-run warfare, and were incapable of maintaining prolonged sieges. The wall was designed to absorb the impetus of surprise attacks and to impose the will of the Chinese defenders on their attackers. It made cavalry warfare impossible and dictated static warfare on terms favourable to those holding the wall.

Its power is still apparent. Following ridges, climbing, and tumbling, it snakes through the landscape like a great work of nature rather than of man.

81

TAQ-I KISRA
Ctesiphon, Iraq | c. 6th century CE

In the early seventh century, Ctesiphon was one of the largest cities in the world. Standing on the banks of the Tigris, it had, in the second century BCE, been the winter capital of the Persian Parthian Empire, and then of the Sassanian. So important was the city because of its wealth and strategic location, upon a rich trade route and major river, that it was constantly a target for Roman attack and occupation. In the second century CE alone it was taken three times, including by Trajan in 116 and Septimius Severus in 197.

Little evidence now survives above ground of all this power, wealth, and glory – apart from one sensational structure. This building is the stuff of legend; it has inspired the fantasies of succeeding rulers in the region and has had a profound influence on Middle Eastern and Central Asian architecture. This is partly because of its sheer size and engineering ambition. Now known as the Taq-i Kisra, it perhaps dates in origin from the time of the Sassanian king Shapur I, who reigned from 241 to 272 CE, with the surviving structure possibly constructed in the fourth century but more likely for Khosrau I, Hormizd IV, or Khosrau II, who together reigned from 531 to 628 CE. What survives are the ruins of an astonishing edifice – a vast, vaulted hall incorporating what remains the widest (25.3 metres/83 ft) and highest (36.7 metres/120.5 ft) single-span unreinforced brick arch ever built.

This was a palace, and the vault, essentially a huge niche, marked the place of entrance or reception – perhaps also of banquets and the location of the throne – and seems to have been inspired by ancient Mesopotamian structures. Comparable central entrance recesses can be found in 2,700-year-old Assyrian palace architecture and in the stone-built palaces dating from the first and second centuries CE in the Parthian city of Hatra, in northern Iraq. The arrangement is also reminiscent of the traditional open-fronted reception tents of nomadic Arab sheiks.

All that now remains of the Ctesiphon palace are the front portion of the arch and flanking elevations ornamented with tiers of blank arcading. This isn't much – and one of the elevations is a twentieth-century reconstruction following a collapse in 1909 when the Tigris flooded – but it's still enough to evoke the wonder of the original building.

Several descriptions of the palace state that it is built of sun-dried bricks, but this has always astonished me. Why would anyone come to this conclusion? Kiln-fired brick had been used in the region since at least 3,200 BCE, and such a demanding structure would surely require the hardest and most stable material available. And how could bricks made of no more than dry mud endure for so long, particularly for the centuries when the structure was an abandoned and unroofed ruin? When I visited Ctesiphon in late 2002, I examined the bricks very closely indeed. If sun-dried, they are the hardest sun-dried bricks I have ever seen. They have all the characteristics of fine kiln-fired brick.

But while the material of construction remains – somewhat surprisingly – debated, the process of construction is pretty well agreed upon. The challenge was to create the huge vault without the need for scaffold or timber centring (wood was scarce), so the arch had to be self-supporting during the construction process. The Sassanians had got this down to a fine art, and in the process managed to construct arches with minimal lateral thrust and of great strength.

The most immediately obvious features of the arch are that it is not semi-circular in shape but parabolic, or an inverted catenary – an immensely strong form – and that the lower portion of the arch is constructed in horizontal courses that rise vertically and only gradually corbel inwards. Such a structure exerts so little outward thrust that stability can be retained by thick walls on each side of the arch without the need for elaborate buttressing.

When Ctesiphon fell to Muslim invaders in 637 CE, the palace was used as a mosque. This may in part explain why the *iwan*, or vaulted arched recess – open on one side and much like the vault at Taq-i Kisra – became a standard feature in mosque design. In the seventh century in the Middle East, the huge brick vault of the Taq-i Kisra was regarded as an architectural wonder – comparable to the magnificent engineering legacy of Rome – and so was one of the great, if now largely overlooked, objects of inspiration and emulation.

82

FORBIDDEN CITY
Beijing, China | 1406–20, with subsequent reconstruction

The Forbidden City is roughly square in form and isolated from Beijing by high walls and a wide moat, and covers 72 hectares (178 acres). It contains more than 800 buildings, virtually all built of timber, making it the world's largest collection of historic, timber-built structures. And the colour spectrum is mellow and memorable – the wood is mostly painted dark red with dull gold-colour trim, combined with magnificent, baked-brick paving and deep orange terracotta roof tiles. Although a city within the city of Beijing, and complex in the mix of uses it contains, the Forbidden City is, in essence, a vast palace, arguably the largest on earth.

The city is one of the great monuments of the Ming dynasty that ruled China from 1368 to 1644. Most of it was built between 1406 and 1420, although much has been rebuilt over the centuries following fighting and fire. This was an exclusive world of imperial power and privilege until the abdication of the Last Emperor, Pu Yi, in 1912, and remains the greatest symbol of the Old Regime. Consequently, the fact that this fragile collection of easily combustible timber structures survived the turbulent decades of Republicanism, Communism, and Cultural Revolution is a miracle. Now, as is the strange way of human affairs, the Forbidden City has been transformed from a hated symbol of imperial rule to a much-loved expression of Chinese history and civilization.

Construction was initiated by Zhu Di, the third Ming emperor, who ruled from 1402 to 1424, and in Chinese the city was, and still is, called the *Zijin Cheng* – the Purple Forbidden City. It's an odd name, until one realizes that 'purple' refers to the North Star, which was thought to be the realm of the celestial emperor. It was presumably the constant and reliable nature of the North Star that encouraged the Chinese to view it as the abode of a divine being – one who created order out of chaos – just as this sense of stellar stability had done for the Egyptians over 4,000 years before. The dominant religious beliefs in China during the Ming period were Buddhism and Taoism, the latter of which in particular venerates nature and believes in celestial 'immortals' dwelling in the mountains and exercising a benign influence over life on earth. To the north of the Forbidden City, an artificial mountain – now in Jingshan Park – was formed from spoil from excavating the moat. It is probable that the Forbidden City and its axially related 'mountain' were created to attract the star-roaming

Taoist 'immortals' and, by constructing a divinely inspired abode for the terrestrial emperor, to bring to earth the attributes of the celestial emperor of the North Star.

So the Forbidden City is, in its plan and buildings, an attempt to create the celestial city – heaven – on earth. Zhu Di chose to express such an audacious claim through the architecture of his new palace because the Ming believed that the Mandate of Heaven – the *t'ien ming*, or divine right to rule – had passed to them. They saw themselves as the dynasty of heaven, and the Forbidden City, intended to proclaim and legitimize this claim, expresses the Mandate in a very precise way.

From the south or Meridian Gate on Tiananmen Square to the North Gate runs a straight route that connects all the main elements of the city and that extends beyond the walls of the city to establish the main north–south axis of Beijing. Within the walls of the Forbidden City this route passes through seven courts, with several halls, and a garden no doubt expressing the seven heavens.

But the Forbidden City is not only a diagram of heaven; it is also, arguably, an analogy of the human body. The straight route can be seen as the spine connecting the private imperial residential quarters in the north portion of the city to the accessible area in the south portion. Of course, the northern portion represents the head and brain, where decisions are made which pass down the 'spine' to the southern portion of the city. The southern portion is the executive realm, where imperial ideas and dictates are given physical and tangible expression. At the centre of the city, between these two realms, are the stupendous Hall of Supreme Harmony – the main reception and ceremonial building – and the Hall of Central Harmony and Hall of Preserving Harmony. Placed on a tiered base, disconcertingly reminiscent of a step pyramid, these halls represent the heart of the Forbidden City.

To the north of these halls, in the most 'forbidden' and private part of the city, is the imperial residential court, with the emperor's Palace of Heavenly Purity astride the main axis and north of it the empress's Palace of Earthly Tranquility. Either side are smaller buildings for the emperor's concubines and members of the royal household. There is also the picturesque Imperial Garden, with its pools and temples, and then the Gate of Divine Might, leading outside to the man-made holy mountain. The fact that the Forbidden City survives is not the only miracle; in a world of increasing cacophony and chaos, it also retains its tranquillity, poise, and harmony, as if it really were a little piece of heaven on earth.

CHAPTER 6

MATERIAL MATTERS

ARCHITECTURE CAN BE INSPIRED by myth, be organized around sacred geometry, embody esoteric symbols, and make statements about pride, power, and national identity. But, ultimately, buildings have to live in the material world – to function and endure they have to be well built. As the Roman architect Vitruvius observed over 2,000 years ago, architecture must possess firmness and commodity as well as delight. Some buildings conceal or deny their material and means of construction, while others express them boldly and honestly. But, whichever approach is followed, the materials from which a building is made and with which it may be clad or decorated are a very significant issue. They can notonly dictate methods of construction and appearance but also reveal much about the people and society that made the building. The choice and use of construction materials can make philosophical or artistic statements, reflect patterns of trade, and make clear the technology and economics of the time of construction.

Timber was used at the early thirteenth-century Barley Barn at Cressing Temple, Essex, as a material with which to construct an engineered timber frame, and at the early eighteenth-century Church of the Transfiguration at Kizhi, Russia, as squared and horizontally laid logs. Marble, stone, paint, and gilding were used for the exquisite and jewel-like fifteenth-century Ca' d'Oro, Venice. Sun-dried brick and mud render were employed at Shibam, Yemen; a shot-proof wall of sand, clay, and rice at the largely seventeenth-century Hameji Castle, Japan; wrought-iron and kiln-fired brick at the Beurs van Berlage in Amsterdam; weather resistant and self-washing glazed earthenware, or majolica, in Viennese apartment blocks built in the late 1890s to the designs of Otto Wagner; shimmering white blocks of chalk at the early-twentieth-century Marshcourt, Hampshire, England. Steel and glass were used in the Farnsworth House, Plano, USA, of 1945–51; steel and glass along with reinforced concrete in British High Tech architecture of the 1970s; and steel, wood, glass, and titanium at Frank Gehry's additions to the Art Gallery of Ontario in Toronto, Canada.

Perhaps most critical of all is the role materials can now play in the current and ever-evolving international architectural campaign to help save the planet from environmental damage and to mitigate global warming. Materials used in construction should involve minimal pollution, or emission of 'greenhouse gases', in their manufacture and transport, should be sustainable and derived only from renewable resources, and offer good insulation from heat and cold. The design should also aim to liberate a building as far as possible from reliance on services fuelled by the consumption of fossil fuels, make the building self-reliant, and allow it to derive energy from the sun and the wind. In addition to all these demands, 'green' architecture should also possess beauty and a poetry of form to ensure that it makes a positive contribution to its setting. A tall order indeed, but the Rozak House, Northern Territory, Australia, suggests that green design can also produce elegant architecture.

83

THE BARLEY BARN
Cressing Temple, Essex, England | c. 1205–35

The Barley Barn was constructed in the very early thirteenth century, according to tree-ring (or dendrochronological) dating of its main structural timbers. This makes it one of the oldest still-standing timber-framed structures in the world. The barn was built – along with the adjoining, larger, more structurally spectacular and fifty-years-later Wheat Barn – for the Knights Templar, who, in around 1140, were given the land on which the barns stand by the Empress Matilda. This pair of barns reveals much about the potential of timber as a building material.

Timber remained a primary material for the construction of houses (and more humble architecture) in the richly forested regions of northern Europe until well into the eighteenth century. Systems of construction varied tremendously, usually reflecting the nature of the timber available, as did the technology of joints. The basic division is between squared or partly shaped staves or logs, used horizontally or vertically to form load-bearing walls, and skeletal, timber-framed construction. The former system evolved where straight but not tight-grained pine or fir was readily available – typically Scandinavia and northern Russia – and the latter where oak was plentiful. Oak does not grow straight and true, but it is immensely strong and tight-grained, making the fabrication of complex joints feasible; and when such joints are possible, and the inherent strength of oak is utilized, strong frames can be wrought and engineered architecture can be achieved.

Oak-framed construction evolved in different parts of Europe in related but distinctly different ways, reflecting powerful regional characteristics. Methods of construction also changed through time, responding not just to the developing technology of construction but also to the availability and quality of building timber. In addition, the location and function of a building had a significant influence on design. Obviously, rural and urban buildings could display fundamental differences from each other, as could houses and barns. For example, houses could have jettied – or projecting – upper floors to increase floor areas, and the corners of jettied buildings could have 'dragon beams', set diagonally to the standard floor joists and beams, helping to support the floor and jetty.

Also important was the artistic attitude to the structural frame. Sometimes it was merely a functional system that would be concealed when the building was complete, perhaps plastered over or covered with boarding or masonry; or the frame could be designed to be exposed, in which case decorative details – not necessary for function – might be added to the frame, and more timber would be used than was needed for solidity in order to make the building more visually pleasing and impressive.

The Barley Barn, at Cressing Temple, Essex, is an early and superlative example of oak-framed construction. This is a building technique that almost defies gravity. Timber stave construction is, like masonry construction, a direct expression of the laws of gravity. But in timber-framed construction strength comes from the materials used and – primarily – from the manner in which the frame is constructed and the skill with which the varied joints are designed, cut, and fastened. The type of joint depends on the type of timbers to be connected, the joint's location in the frame, the job to be done, and the force to be resisted. In most timber frames the trickiest forces to be contended with are not to do with vertical loads being transferred down, but with lateral loads due to wind pressure. To counter these loads, different systems of braces were evolved, usually curving or set on the diagonal, to strengthen corners and to connect horizontal and vertical timbers.

The Barley Barn is 36 metres (118 ft) long, but seems to have been shortened in 1420 and its roof rebuilt in the sixteenth century. Typical of large medieval barns, its wide span – required for functional reasons – is achieved by means of two interior rows of parallel arcade posts that help support the wide roof trusses. These posts divide the interior, church-fashion, into a central 'nave' flanked by aisles. Crossing the 'nave', in the manner of transepts in a church, is the 'threshing floor', where barley grain was separated from the plant by flailing. At each end of the threshing floor are facing doors. These allowed barley plants to be brought in at one side and the grain removed at the other. They also, when the grain was winnowed, created a through draught that helped to separate out the chaff. A great barn is, indeed, a finely honed agricultural machine that, in design, is wonderfully fit for purpose.

The roof trusses comprise tie beams from which rise 'crown posts' that contribute to the support of the collar beams and purlins. The purlins – running parallel with the nave – help in turn to support the rafters forming the roof. As in skeletal, masonry-built Gothic architecture – with its related ribs, piers, and buttresses – timber-framed construction is an integrated system of great sophistication. The outer walls of the Barley Barn are formed with timber studs, making this essentially an all-timber structure.

The visual and structural connections between medieval barns and churches are significant, and in a most moving way confirm the importance of the barn in a medieval rural Christian community. It was a place of vital communal work, where the wonder of God could be venerated through His life-giving works of nature.

The great former monastic barns of north and west France are arguably unprecedented in their scale, majesty of construction, and architectural ambition. Some, like the early thirteenth-century barn of the former Benedictine priory at Perrières, Calvados, have stone-built exteriors and stone-built interior arcades including carved capitals and pointed arches. Others were of huge scale – the Cistercian monastery of Clairvaux once owned a seven-aisled barn on its grange at Ultra Albam, while a five-aisled barn (or central nave flanked by two pairs of aisles) survives at Parçay-Meslay, Indre-et-Loire. Built in the early thirteenth century, and the last of its kind, the latter has a masonry exterior of some architectural ambition.

These French barns display with clarity the principles that ennoble timber-framed medieval barns. They show that architectural beauty comes from the honest expression of the means and material of construction and from the quest for functional perfection. Significantly, architectural adornment is limited to the masonry elements, while the timber portions retain their practical purity and almost sacred simplicity.

84

CA' D'ORO
**Grand Canal, Venice | c.1428–30 | Marino Contarini,
Giovanni and Bartolomeo Bon**

Ca' d'Oro is, in a sense, the epitome of the fifteenth-century Venetian palazzo. Sited magnificently on the Grand Canal, near the Rialto Bridge, it embodies the true spirit of Venice's own and distinct brand of Gothic – a heady mix of familiar northern European forms and motifs with Eastern and Islamic influences, a result of Venice's extensive trade and cultural connections with the Ottoman Empire.

The most notable element of the main elevation is the tiers of loggias, each serving a different community or function within the building. The internal organization of Ca' d'Oro is typical of many of the larger palaces on the main canals. As a rule these palazzi served not just as homes but also as places of business or trade, or even partly as warehouses; consequently, ground floors were given over to the functional requirements of receiving people or goods by boat, with a loggia on the canal forming a landing stage and leading to a large and open courtyard. From this courtyard a grand staircase usually led to the main upper rooms. If a small canal ran along one side of the palazzo, that would usually be the route that would deliver goods to a side entrance. First-floor apartments were used for formal receptions and entertainment, with second-floor apartments reserved for family use.

Ca' d'Oro was built from about 1428 to 1430 for Marino Contarini, using sculptors Giovanni and Bartolomeo Bon as architects. Contarini was the procurator of San Marco, came from an established family, had married well, and wanted to make his mark in the city with a striking palace on one of the most dominant sites on the Grand Canal. The design he orchestrated is a subtle blend of fashion and novelty, with conservative and traditional details intended to convey patrician pedigree. For example, the ground-floor waterside arcade is in the old Veneto-Byzantine tradition – possibly retained from the earlier building on the site – and the asymmetry of the main, canal-side elevation seems to have been intended to evoke memories of Torresella; these are flanking towers that were used to give dignity and prestige to Veneto-Byzantine palaces. At Ca' d'Oro, however, a tower was never built but was seemingly implied by an area of wall that marks not the base of a tower but merely the location of comfortable and convenient rooms set off the *portego*, or traditional gallery-like hall.

In contrast to these references to past details and forms, the first- and second-floor loggias are cutting-edge Gothic. Their ornate tracery establishes a model for palazzi in later fifteenth-century Venice and is a refined and more domestic example of comparable traceried designs found in the slightly earlier Doge's Palace.

The main façade is now clad with much-mellowed stone but originally would have been an astonishing sight, and probably not much to today's tastes. In 1431 Contarini had carvings picked out in ultramarine and red, and had gilded whatever was not painted or marble-clad. This is what gave the building its name – the 'golden house'.

By the late fifteenth century the ownership of Ca' d'Oro had moved, by marriage, from the Contarini family, and it changed hands many times until in 1840 it was bought by the Russian prince Alexander Trubetsko and given to the ballerina Maria Taglioni. Unfortunately she turned out to be foolish and vain and demolished the famed 1420s courtyard staircase (since reconstructed in partly authentic manner) and first-floor balconies.

It was at this time that the English art historian and polemicist John Ruskin got to know the house. He was in Venice compiling his magisterial work *The Stones of Venice*, published between 1851 and 1853, which sought not just to catalogue and categorize Venetian Gothic architecture but also to stop the demolition of fine examples and to promote sensitive repair. Ruskin's writings on the Gothic became highly influential. He analysed it under various headings that attempted to explain its qualities – 'savageness ... changefulness ... naturalism' – and celebrated Gothic's artistic freedom and air of invention. His profound study captured the public's imagination and had ramifications that were perhaps surprising. At a time when Gothic was being promoted in Britain as the basis for a modern national style, Ruskin's writings encouraged a preference for aspects of Venetian Gothic. This style was admired notably for the way in which buildings were given colour and character through the mix and exposure of different natural materials – bricks, stones, and marbles – that were used in structural as well as ornamental roles. The architectural term for the consequence of this mix was 'structural polychromy', and along with Venetian Gothic forms and details, it became a key part of the later nineteenth-century Gothic Revival in Britain.

In his book Ruskin referred to Ca' d'Oro as an exemplar: 'A noble pile of very quaint Gothic, once superb', with 'fantastic window traceries' and 'the capitals of the windows in the upper story, most glorious sculpture'. But for Ruskin the building was tainted by tragedy: 'I saw the beautiful slabs of red marble which formed the bases of its balconies … dashed to pieces when I was last in Venice [and] its glorious interior staircase, by far the most interesting Gothic monument of its kind in Venice … carried away … and sold for waste.'

85

SHIBAM
Yemen | 16th century

Shibam is a combination of extraordinary things. It is a city with an ancient history – stretching back perhaps 2,500 years – in a region characterized by non-urban nomadic culture. And it sits within an oasis on the banks of a watercourse yet is built, seemingly most precariously, of sun-dried mud bricks. Given the vulnerability of this material to moisture, it is little short of miraculous that many of the tall houses in the city date, in origin or at least in part, from the sixteenth century.

The peculiar nature of Shibam is explained by the circumstances in which it was created and in which it now exists. The nomads of the region were traders, and Shibam seems to have been established as a stable and permanent centre of merchant endeavour where goods could be gathered and safely stored, and where they could be displayed, bought, or bartered.

And its riverside location – potentially catastrophic for a city made of no more than dried mud and a few palm tree trunks – is explained by a series of subtle but very significant factors. For most of the year the river is nothing more than dry and sun-scorched riverbed – a *wadi* – and when its fills with fast-flowing water during the brief rainy seasons it carries alluvial earth that is the very lifeblood of Shibam. The annually arriving deposits of rich earth ensure that brick-making is possible – and sustainable – and, more importantly, enriches the soil of the oasis that surrounds the city. The precious nature of this oasis, and the threat of inundation, means that the city has, for millennia, perched on a slightly raised rocky plateau located next to the river within the heart of the oasis. Provided that houses are built on this slightly raised land, they are generally safe; to build lower would not only make buildings vulnerable to flood but would also destroy valuable agricultural land.

As the city thrived, its population grew, but given the precariousness of its situation, how was it to expand? Upwards, of course. So Shibam is, predictably, known as the 'Manhattan of the Desert', with its families living in tower houses generally five or six storeys high but some as tall as eleven storeys. Rather than being free-standing, these towers generally abut one another, like terrace houses, and are grouped to create inner 'service' courts into which sewage and waste were originally discharged. Presumably, when this pattern of development was first established, the individual tower houses were occupied by the region's leading families or clans, each of which expressed its wealth, power, and status through the size and ornamentation of its tower, and the tower's location within the city. Key locations, places of pride and display, included the sites on the edge of the city where tall and abutting towers, incorporating terraces, form what is effectively a city wall with fighting platforms.

In their construction and internal organization the towers are highly functional, designed not only to be defensible but also to maximize the insulation qualities of clay construction, utilize traditional techniques of natural ventilation, and offer space for storage and living.

I went to Shibam in 2005, in what was, as then seemed likely, a rare window of opportunity to travel freely in this troubled region. When there, I was invited inside one of the taller, older, and better-preserved tower houses by a family that had resided in Shibam for generations. I asked them the age of their house. It was difficult to date, they answered, perhaps 250 years old. But since it is made of dried mud, they pointed out, the house needs constant maintenance, with large portions regularly rebuilt. Neglect for only a few years, if the rainy seasons have been harsh, can lead to collapse.

This house followed the traditional arrangement. There was a fortified entrance and storeroom and stabling on the lower couple of floors, and an internal top-lit court – shared with neighbouring houses – to allow air and light into the heart of the house. The walls were thick, battered at their base and lime-washed at high level for additional weather protection. On the second floor was the first main living room, with large windows, shaded with lattice grilles; above these were small openings through which, as part of the system of natural ventilation, hot air exits, allowing cooler air to be sucked in through the windows below. This large room, with its ceiling supported by four ornamental timber columns, was for the men of the house. A similar room, with adjoining kitchen, was located above for the women, then bedrooms and roof terraces.

This tower house is an admirable machine for living in and, like the other traditional houses of Shibam, visually abstract – indeed modern – in appearance. This is an architecture determined by functional and utilitarian demands, by a resolve to push the potential of high-rise mud construction to the limits.

Since my visit, Shibam has been hit by abnormally severe flooding, which in 2008 damaged the ground floors of some of the tower houses, and in 2009 tourists in the city were targeted by Al Qaeda terrorists and four people were killed. Despite being a UNESCO World Heritage Site, Shibam's long-term future currently looks bleak.

86

HIMEJI CASTLE
Himeji, Hyogo Prefecture, Japan | 1333–46, 1581, 1601–09, 1617–19

Himeji Castle is one of the finest surviving examples of military architecture in Japan. It was designed, from the mid-fourteenth century, as a highly efficient and practical military machine offering complex systems of defence in depth. Yet as with so much essentially functional Japanese architecture, the hill-top fortress possesses a delicacy of detail, fineness of form, and picturesqueness of profile that, from a distance, makes it look more like a fairy-tale palace. Here beauty and stern purpose appear to be reconciled in a truly remarkable and memorable manner. Without doubt Himeji is one of the most poetic castles in the world, and much of its strength and ornament is derived from the fine and judicious working of its material of construction.

The castle was first completed in 1346 but was completely rebuilt and enlarged between 1601 and 1609 for the *daimyo*, or lord, Ikeda Terumasa, with several additional defences added from 1617 to 1619. It is the castle created then that largely survives, still in remarkably intact and authentic condition.

What you now see when first arriving below the castle is a multi-storeyed and multi-roofed keep that was originally a mighty redoubt set within a complex system of defence that involved numerous curtain walls, courts, and moats, as well as the city itself, which was also surrounded by walls and moat.

For eyes accustomed to the mighty stone walls and towers of European and Middle Eastern castles, Himeji appears not only rather pretty but also somewhat fragile. The lower portions of its outer walls are wrought of stone blocks, rising in visually graceful curves but steep enough to prevent them being scaled. And on these stone bases sit delicate white-painted pavilions, with roofs made of timber and tiles and walls seemingly of lathe and plaster.

Warfare in late sixteenth- and early seventeenth-century Japan was highly ritualized and dominated by tradition. As in most other aspects of Japanese life, technical innovations, if not forbidden, were kept under strict control. Ideas and new weapons did appear in Japan from the outside world but were applied in a controlled manner. This is particularly true of gunpowder-based weapons, which were kept subservient to traditional weapons within a code of battle dominated by the sword and the concept of honourable single combat. Consequently weapons technology in Japan soon became fossilized, with early forms of firearms such as matchlocks still being used when long superseded in other parts of the world, and certain types of heavy artillery, such as large-calibre and long-range culverins or cannons, never coming into favour at all.

Consequently, Himeji Castle was created for a particular type of formalized warfare. So at Himeji there are no thick-walled casemates or fortified platforms for the location of heavy defensive and offensive artillery. There was also no need to build defences against such powerful weapons, so roofs could be merely tiled as they did not have to withstand plunging shot, and walls could be relatively slight, although the seemingly plaster-made white-painted walls of the upper works at Himeji are in fact very thick and are made of a tough composition of sand and clay bound with rice glue that is impenetrable to matchlock balls.

The subtle and particular nature of Himeji's system of defence can best be appreciated by walking from the ground below its wall to its main keep – the route that any successful attacker would have to negotiate. Like contemporary castles in other parts of the world, the secret of Himeji's strength is its concentric rings of defences culminating in the tall, central keep or redoubt. All is based on the principle of defence in depth, with walls and ditches complementing and protecting each other so that the deeper the attacker penetrates the defences, the more intense and focused is the fire that can be brought to bear upon him.

First there is an outer moat that would have to be crossed, then the strongly fortified gatehouse, and beyond that the outer court or bailey. To have reached here an attacker would have had to weather fire from loopholes placed at regular distances along the parapet of the curtain wall. These loopholes are of different forms – triangular, rectangular, and round – which give them an abstract beauty, but all was based on function, for some loopholes were designed for use with matchlocks and others for use with bows.

The attacker would now have been trapped in a killing ground, with missiles raining down on him. Beyond, towards the heart of the castle, is another well-fortified area defended by high-level loopholed walls. Beyond this inner defence is the ultimate goal of an attacker – the towering keep. Here there is an element of psychological warfare. The gate leading to the court immediately below the keep appears hardly fortified, but this is to lure the attacker in, because immediately above the gate itself is a secret trap door – a frightful 'murder hole' through which missiles or boiling liquids could be used to slaughter attackers below. The doors into the keep itself are also protected by 'murder holes', and inside, this redoubt is full of defensive tricks. Its plan is labyrinthine to confuse and disorient attackers, and in the four corners of each main floor are small strong points – miniature castles – in which a few warriors could lock themselves and fire into the interior of the castle through loopholes at attackers.

Not only was Himeji Castle never taken by assault, it was never even attacked. Its strength and cunning design were deterrent enough. The castle is a sublime monument to the

Japanese art of war, in which – as exemplified by the Japanese sword – utility and beauty work together with a thrilling harmony and perfection of purpose.

87

CHURCH OF THE TRANSFIGURATION
Kizhi, Republic of Karelia, Russia | 1714

The Church of the Transfiguration at Kizhi, Russia, is a moving masterpiece of timber construction. It is made of pine logs that are laid horizontally, and accurately shaped by axe and adze so joints are tight, with no filling required to keep the interior windproof. One of the secrets of the longevity of the softwood logs forming the church – amazing, given the wet climate – is the technique of construction. A saw, cutting through the grain, can weaken the log because such a cut allows water to penetrate into the heart of the timber. But an axe and an adze, cutting with the grain, expose surfaces that are more impervious to water penetration. And the logs are beautifully jointed; at some corners they neatly overlap, as in a log cabin, while at others the ends of the abutting logs form lap joints that are cut flush.

The church rises 37 metres (120 ft) high, its form a cascade of cross-topped, timber-built, and timber-clad onion domes. It is part of a group of sacred timber-built structures, including a bell tower, enclosed within a fortified log-built stockade and burial ground. All within is consecrated, a spiritual refuge called a *pogost*. The plan of the church is formed by two squares of equal size, one turned at 45 degrees over the other to form an eight-pointed star – a favourite symbol in early Christian churches as well as in mosques. The inner space defined by this turning of one square over another is a regular octagon, and it's this form that is the key motif of the design. On alternate faces of the octagonal plan are placed square extensions to create a cross with four arms of equal length – a Greek cross.

The church is four storeys tall, and as each level is smaller in area than the level below, it has a tiered, pyramid-like profile. This structure boasts no fewer than twenty-two domes, each an image of the vault of heaven, and loaded with symbolism. The large dome at the top and the four around and slightly below it are a standard design in Russian Orthodox Churches and represent Christ and the Four Evangelists – Matthew, Mark, Luke, and John. And the number of domes is probably a reference to the Book of Revelation, a most important text for the Orthodox Church, which is composed in twenty-two chapters.

So, like many churches, this structure is a prayer, a sacred text, and a spiritual message made manifest.

88

BEURS VAN BERLAGE
**Damrak, Amsterdam, The Netherlands | 1896–1903 |
Hendrik Petrus Berlage**

The Beurs van Berlage, or Berlage Stock Exchange, is a building that performs on many levels and carries many meanings. It was consciously built to reflect 'the resolute and practical spirit of Amsterdam's merchant class', and its architect, Hendrik Petrus Berlage, wanted the design to be a bridge between history and Modernism and to define his quest for an architecture that, though rooted in the past, would be an appropriate harbinger of the coming century. These concerns made Berlage a crucial intermediary in the Netherlands between traditional design inspired by history and functional Modernism, and for this inspirational role he is now considered the nation's 'father of Modern architecture'. The Exchange is Berlage's first major exploration of fusing old and new in an original and creative manner.

Berlage also recognized that, given its importance for the mercantile community of Amsterdam, the building not only had to appear as 'a palace of commerce and finance' but also had to contain a programme of edifying and symbolic art, often signifying the varied functions taking place in the building. To identify themes and forms Berlage enlisted the help of the poet Albert Verwey, who, together with the body of artists to be engaged on the project, resolved to make the building, in its symbolic ornament, a *Gemeenschapskunst*, or an expression of the 'art of the community'.

Berlage was forty years old when, following an abortive competition, he was given the daunting commission to design the Exchange. He had studied in Zurich under the German classicist Gottfried Semper, then travelled Europe to experience exemplary historical buildings at first hand. In 1896 he accepted the challenge to create a building that was complex and functionally demanding, and that would, through its character and symbolism, realize the dreams and reflect the aspirations of Amsterdam's mercantile community. Berlage was at liberty to choose any style for the building. His approach was logical, painstaking and open to scrutiny. At a public debate organized by the municipal council, he presented his initial design and explained the referencesto historic and modern prototypes. During the next two years Berlage produced three different versions of the scheme, each one part of a continuing quest for rational simplification and economy of means aimed at realizing an 'honest synthesis' between aims, methods, and inspirations. Gradually, super-fluous and time-bound historic details disappeared. For exam-ple, a corner tower – a seemingly inescapable traditional expression of mercantile pride and urban significance – was gradually whittled down, lost its spire, its turret, and its loggia, and by the time construction started in 1898 was a simple, streamlined and proto-modern affair, to a degree anticipating the stripped-back Modernist tower at Hilversum Town Hall, near Amsterdam, designed in 1928 by Willem Marinus Dudok.

Ultimately, with the Beurs, Berlage achieved a fascinating fusion between a free-form version of Dutch traditional brick-built architecture, in which his historic details are gener-ally Gothic in inspiration, and iron-made engineered construction anticipating Modernism. Most impressive are the roofs of the building's high and top-lit main exchanges – for commodities, for stock, and for grain – and in the top-lit passages that link the main spaces. These incorporate wrought-iron lattice trusses of various scales and forms. Those in the commodities exchange are arched in form, with each truss rising from a delicate pin-joint set on a stone corbel placed atop an elemental brick buttress. The ironwork is unembellished and a bold expression of the materials and techniques of construction, and works in happy, if contrasting, harmony with the brick-wrought inner walls of the exchange. These, in design and construction, are evidently inspired by history but possess a refined and sculptural simplicity that matches to perfection the refined character of the iron.

Within a decade of the opening of the Beurs in 1903, Berlage's career had taken a new direction. In 1911 he travelled for two months in the United States and saw the works of Henry Hobson Richardson, characterized by their massive neo-Romanesque forms; of Louis Sullivan, who had pioneered high-rise construction and rethought the role of ornament in architecture; and, most importantly, of Frank Lloyd Wright. Berlage felt a great affinity with Wright's architecture. He saw within it an acceptance of the rise of an industrial society, and he shared what he perceived to be Wright's belief that the expression of modern technology was the hallmark of an enlightened and progressive democratic society. For Berlage, Wright was 'a man who has freed himself from tradition without severing his ties with the past, who does not imitate the past but understands it as a historic phenomenon'.[1] These perceptions did not transform Berlage's architecture or architectural ideas; rather, they confirmed them. From the beginning of his career he had sought architectural change through evolution, not revolution, and a creative continuity between the past and the present. But the power of the American experience did, as has been pointed out, mark 'a turning point' in Berlage's career.[2] This is made clear in his later works, which reached a tremendous climax with the rational and simple brick-built Modernism of his last work, the Municipal Museum, designed in 1927 and completed in 1935 in The Hague.

89

THE MAJOLIKAHAUS
Vienna, Austria | 1898 | Otto Wagner

In 1894, during his induction speech as a professor at the Viennese Academy of Fine Arts, Otto Wagner offered as his architectural creed a conviction developed from the theories of his hero, the German architect Gottfried Semper: 'Art and artists should and must represent their times. Our future salvation cannot consist of mimicking all the stylistic tendencies that occurred during the last decades ... Art in its nascence must be imbued by the realism of our times.'[3] Wagner's contemporaries such as US architect Louis Sullivan could contemplate a modern architecture in which buildings 'quite devoid of ornament' could 'convey a noble and dignified sentiment' and be 'comely in the nude', while the following generation – notably Adolf Loos – would condemn ornament as a 'crime'. But for Wagner, ornament in architecture was not passé or a crime. It was fundamental. What was wrong in his view was to load contemporary buildings – often unprecedented in their scale, function, and means of construction – with ornament evolved by past cultures for very different types of buildings.

Wagner, in his 1896 book *Modern Architecture*, argued that architects must use new materials and forms in architecture to reflect the way society was changing. He also acknowledged the time-honoured functionalist dictate that necessity is the sole mistress of the art of architecture. But this did not mean the end of ornament – far from it. For Wagner it was neither seemly nor civilized to take the functionalist theory to its logical conclusion and allow modern architecture to be entirely liberated from surface, and non-functional, decoration. For Wagner that would have been tantamount to barbarism, to strew the streets of Vienna with ugly utilitarian industrial-style constructions that would lack visual poetry and artistry and that would fail to delight and elevate. So for Wagner the question was not whether or not to ornament architecture but what form contemporary architectural ornament should take, and the degree to which it should be inspired by contemporary technology and artistic sensibilities.

These questions were tackled, if not entirely answered, by a group of artists and architects who in 1897 formed the Vienna Secession. The aim of the group, which had the artist Gustav Klimt as its first president, was to liberate art from dowdy history-inspired ornament imposed by the conservative state-sponsored Vienna *Künstlerhaus*, and replace it with a style of ornament freed from direct historical influence and evolved by free-thinking artists. So the neo-classical, the neo-Gothic, and the neo-Romanesque were out, and the quest was on for a style that lived up to the motto emblazoned above the door of the 1897 Secession Building, designed by Joseph Maria Olbrich: 'To every age its art. To art its freedom.'

The artistic achievements of the Secession can now seem pretty tame when compared with the revolutionary transformations imposed a few years later by ruthless Modernists who swept all ornament away, along with historicism. The success of the Secessionists' escape from historicism is even debatable. Olbrich's Secession Building does not sport conventional classical orders, but it is domed and has corner pilaster strips and a rudimentary cornice so it feels classical, if only in an elemental way. In painting, success was more marked, as Klimt in particular embarked on a modern quest delving into the dark world of the subconscious and the instinctive, echoing the writings of Vienna contemporary Sigmund Freud.

Otto Wagner was not a founder-member of the Secession, but his work expresses best its architectural potential. Notable are the apartment blocks of 1898 at Linke Wienzeile, Vienna, that feature and frame his remarkable Majolikahaus. The group of apartments forms a dominant urban feature, with their elevations a dazzling display of the new style of ornament, liberated from direct reference to historic precedent. As with Louis Sullivan's contemporary spectrum of ornament, the inspiration for Wagner's decoration is derived largely from natural forms – flowers, tendrils, palm leaves – and in its lush, organic, and asymmetrical form is, with hindsight, now classed as one of the more characterful expression of the worldwide Art Nouveau movement of the 1890s.

Another influence was the Scottish architect Charles Rennie Mackintosh, whose work – notably the Cranston Tearoom in Glasgow of 1897 – had been widely discussed and had made a major impact on Viennese designers. The painter Friedrich Ahlers-Hesterman said the Cranston Tearoom was 'like a dream', with its narrow panels, grey silk, and 'little cupboards ... straight, white, and serious looking, as if they were young girls ready to go to their first Holy Communion'.[4]

More striking than the motifs used by Wagner at the Majolikahaus are the scale, and the integration of ornament with new technology. The floral forms cover the entire face of the Majolikahaus, growing asymmetrically, as it were, from right to left. The ornament is realized in glazed earthenware, or majolica, a robust material that is self-washing in the rain and ideal for the building's urban site.

Having achieved this exquisite monument to an artistic moment in time, Wagner moved on, and his 1903 Post Office Savings Bank building in Vienna marked a key phase in the evolution of functionalist architecture. The Art Nouveau excess of the Majolikahaus elevation was replaced by the ornamental expression of the techniques of construction, with marble cladding panels embellished with large bolts that fix them to the structure. The effect is original – and brilliant.

90

MARSHCOURT

Stockbridge, Hampshire, England | 1901–04 (extended 1924) | Sir Edwin Lutyens

Edwin Lutyens is one of the most fascinating British architects. His work is distinguished by clever and careful planning and skilful design, but also by wit, a lightness of touch, originality, and an inspired mix of historic references. His work also commemorates a dramatic and still somewhat perplexing change of taste that overtook British architecture during the decade after 1900. Indeed, Lutyens' work can be seen as a bridge between fundamentally different architectural worlds, in which atmosphere and style were singularly opposed.

The first couple of years of the new century saw the continuation of the Gothic Revival, expressed most successfully through the subtle, sophisticated, and comfortable domestic architecture of the Arts and Crafts movement. Here the sources were largely Gothic, although references were also often made to robust Elizabethan and Jacobean vernacular classicism, to aspects of Christopher Wren's architecture, and to varied seventeenth-century sources united by the curious term 'Queen Anne Revival'. But despite being a strange brew of Gothic and various types of classicism, the Arts and Crafts movement was thoroughly Gothic in its essential spirit. It reflected William Morris' notion that great architecture of the past – primarily Gothic – was hand-wrought in original manner by craftsmen, unlike classical architecture, which, he argued, was usually the result of repetitious drudgery on the part of builders disengaged from the creative process. Arts and Crafts architecture also featured the use of regional and traditional building materials and displayed a Gothic freedom of composition, favouring asymmetrical designs in which internal functions determined external appearances. Then, in around 1905, something strange happened. After decades of dominance, Gothic Revivalism – seemingly deep-rooted, informed by the polemic and theory of such figures as Pugin and Ruskin, and the basis of the accepted national style – was challenged by a return to classicism. Lutyens' work captures this moment of change with uncanny precision.

Marshcourt is the last, and arguably the best, of Lutyens' series of significant Arts and Crafts houses that started in 1897 with Munstead Wood, Surrey, which was also his first collaboration with the inspired garden designer Gertrude Jekyll. Marshcourt is designed in the Gothic spirit and incorporates eclectic historical references including Elizabethan and the late seventeenth-century Baroque of Wren, but it uses these traditional models – in a most brilliant manner – to create a modern home. It anticipates many of those characteristics embraced by pioneering European Modernists such a Adolf Loos, Gerrit Rietveld, and then Le Corbusier – for example, open-plan, split levels, and large windows and *fenêtre longueur* to let light flood in and to offer stunning prospects. Because of this, it is possible to see Lutyens as, in his idiosyncratic way, a forerunner of the modern age. It is also possible to see Arts and Crafts buildings as representing a direction that modern British architecture, informed by history and not averse to decoration, could have gone in if this gentle and evolutionary approach had not been overwhelmed in the interwar years by the revolutionary machine-age Modernism of the so called International Style.

Marshcourt has a fascinating free-form plan, inspired by English sixteenth-century E-plan houses, in which spaces flow together but within which are created dramatic axial vistas. There is also a system of corridors to ensure privacy and convenience. The most striking interior volume is the central, double-height hall that at once evokes memories of medieval great halls and, bathed with light from huge windows, has the feel of a very modern interior. One of the wings contains a billiard room that is set at a lower level to adjoining rooms and thus reflects the building's direct response to its undulating site, but which also anticipates the split-level living soon to be advocated by Loos. Externally, Marshcourt is visually striking. It is faced with blocks of white chalk and flint – used as nodules and knapped – that were extracted when the house's foundations and flanking terraces were constructed. Using local materials to root a building into its site and landscape was a basic Arts and Crafts tenet.

In its form Marshcourt was designed to exploit the site (greatly enhanced by Jekyll), perched on the edge of a plateau and enjoying splendid prospects. The exterior detailing of the house is broadly Tudor Gothic, with much use made of large mullioned and transomed windows that are modern in their simple forms and lack of historicist details. Particularly striking is the long horizontal range of windows on the entrance front that anticipates the *fenêtre longueur* of Le Corbusier.

But even as this Arts and Crafts masterpiece was nearing completion, Luytens had turned his hand to classicism with his 1903 design for the Harriman House in New York State; the neo-Georgian Nashdom in Taplow, Buckinghamshire, of 1905; and the massive, neo-Renaissance Heathcote in Ilkley, Yorkshire, of 1906. Fascinating as these houses are, however, the magic of Marshcourt, with its glowing white elevations and flowing 'Gothic' plan, is absent.

91

ART GALLERY OF ONTARIO
Toronto, Canada | 1910–2008 | Pearson and Darling, Frank Gehry

Founded in 1910, the Art Gallery of Ontario (AGO) is something of a demonstration model of different phases and fashions of museum architecture. It moved from neo-classical Beaux Arts to sleek Modernism and ironic Post-Modernism, and from 2006 was transformed and extended in dramatic and flamboyant manner by Toronto-born architect Frank Gehry.

Gehry is best known for his 1997 Guggenheim Museum in Bilbao, Spain, where he evolved a powerful and distinctive – if idiosyncratic – architecture that glories in combining sweeping organic forms and details, often realized in a range of modern materials. The cladding and external materials used – glass and titanium with limestone – are visually stunning but also, by their nature, of fundamental importance to the design, for they make the previously impossible attainable; for example, the sparkling, billowing organic form at Bilbao, where walls flow and roofs tip and undulate. This is formed largely by glass and titanium panels that reflect light so that the building shimmers in sunlight and allow daylight to help illuminate and enliven the vast galleries.

Gehry's work at the AGO develops and refines some of the key aspect of his groundbreaking and internationally lauded Bilbao design and, once again, demonstrates how much materials matter.

The AGO opened in 1913 within the Grange, a small country house of 1817 that had been bequeathed to the gallery in 1910. In 1916 new galleries were added in a solid Beaux Arts classical style that derived from nineteenth-century Paris and had become very much the house style for public buildings in North America – for example, the splendid New York Public library of 1902–11 by Carrère & Hastings, Grand Central Terminal in New York of 1903–11 by Warren & Wetmore and Reed & Stem, and many projects by US architects McKim, Mead & White. There were numerous extensions and alterations made to the AGO during the twentieth century, culminating in 1992 with the design of a new wing in minimal, brisk, late Modern manner by Barton Myers and Kuwabara Payne McKenna Blumberg Architects.

The AGO had become something of a charming hodge-podge of a building, but this was to change when in 2000 a generous and inspirational Toronto-born art collector and benefactor appeared on the scene. His name was Kenneth Thomson (the second Baron Thomson of Fleet). He had inherited and greatly enhanced a fortune made through broadcasting, publishing, and newspapers, and, towards the end of his life, wanted to give his home city a world-class art museum, with collections spanning many centuries and geographic locations, within a piece of enhanced and world-class architecture. As Matthew Teitelbaum, the director of the AGO, explained in 2009, Thomson believed that much of the art collection he had built up over the decades 'should be shared with the public, that it should be on view at the Art Gallery of Ontario, and that to show both his collection and the existing Gallery collection to advantage would require additional space' so that the AGO could fulfil its promise as the 'centre of Toronto's cultural life, and become more universally recognized as an art institution of international stature'.[5]

With the civic pride, largesse, and cultural sophistication of a Renaissance prince, Thomson endowed the museum not only with many of his prime works of art and $20 million but also with $50 million towards the cost of adapting and extending the gallery. But how was this to be achieved, given that the site had already been largely built upon? And who was to be the architect?

The Guggenheim had worked its magic. The AGO director suggested that Gehry was the man to transform the gallery in the required way, and he and Thomson flew south to Los Angeles to meet Gehry in his Santa Monica office. Patron and architect hit it off, and in late 2002 a project was commissioned to create a Gehry gallery in Toronto, bigger and better than the existing structure.

Complete site clearance was never an option – Gehry was to change the AGO rather than rebuild it – but something had to go. Most boldly, what eventually went was the 1990s wing along Dundas Street.

Gehry started the design process in characteristic manner: he explored the existing building, its collection, and that of Thomson, and then he started to produce a series of exploratory and wiry sketches, feeling his way into the design and letting forms evolve.

The vocabulary of Gehry's building became 'wood and light', with the Dundas Street wing replaced by an undulating elevation, clad largely with glass over curved timber ribs reminiscent of ship construction or a vast mammalian skeleton. The exterior of this new 'Galleria Italia' is striking, clearly a child of Bilbao adapted to circumstances, while inside more space has been achieved. Additional space was created by rationalizing arrangements in the Beaux Arts galleries, through which curvaceous ramps and staircases thread to achieve memorable visual effects. Gehry's trademark use of reflective titanium cladding has been reserved for the elevation to Grange Park, where, in somewhat surreal manner, the gallery rears up as a blue-tinted, shiny box behind the Grange of 1817.

Art-viewing space within the AGO has been increased by 47 per cent to make it one of the largest art galleries in North America. But the real success of the Gehry's design – the blue box notwithstanding – is its relative reticence. The overwhelmingly strident note of Bilbao's sparkling, organic form and self-proclamation has been tamed, with design tailored more respectfully to context and to a city-centre site with a strongly established character. This is a sophisticated design – personal, yet with traditional urban qualities – and is in its way a quiet, indeed gentle masterpiece. It represents a coming-of-age for Gehry and for late Modernism and will surely stand the test of time more certainly than some of his more famed, more self-assertive creations.

92

FARNSWORTH HOUSE
Plano, Illinois, USA | 1945–51 | Ludwig Mies van der Rohe

For those inclined to the minimal vision of the house, this is the perfect home. It is, visually and broadly, little more than a series of sheets of glass set between a thin horizontal flat roof and an equally thin deck, raised slightly above ground level on slender steel posts. The roof and deck, visually equal in depth, are separated by slender steel columns of I-section shape that are extensions of the steel posts supporting the deck. These columns, and the fact that the glass sheets forming the elevations are generally almost square in shape, hint to the initiated that Mies van der Rohe is playing a high cultural game. The house is simple, yet it is reflecting the tenets of classicism – the square is a primary proportion in classical design, so the panes of glass suggest that in this composition classical harmonies are all-important, with the I-section columns evoking the classical orders. Evidently this house, apparently a ruthlessly functional and almost mechanistic diagram of a building, was in Mies' conception an elemental temple-like construction touching on the essence – almost the origin – of classical architecture.

The plan is formed by two rectangular platforms, side by side, that put one in mind of the often enigmatic relationship between rectangular-shaped temples on the Acropolis in Athens. One platform, the slightly higher, is framed by glass to form the house. The other, slightly smaller, is the deck or terrace.

The house is generally open-plan, with just a couple of slivers of wall indulging the modesty of those using the lavatories and bathrooms placed at the centre of one of the rectangles. The views out from the houses are open and magnificent, as – of course – are views in, with any required privacy being achieved by blinds and curtains.

Mies designed the house as an almost theoretical expression of an ideal building – one in which the aesthetic is based on materials and methods of construction yet which is also, simultaneously, rooted in history and the classical tradition. The house can be seen as the final and logical conclusion of Mies' conviction that 'less is more', and also demonstrates that by 'less' Mies did not mean something simple, but a simplification of something very complex – in this case, the history of classical design.

Mies was only able to make his theories tangible due to the sense of adventure, humour, and wealth of his client, Edith Farnsworth, who clearly found Mies attractive and was happy to buy the land and pay the money to build the dream of the genius. Feeling himself liberated from most of the dull considerations that can so easily compromise an architect's muse – like the need of the client to actually live or work in a building – Mies stripped everything back to the essential idea.

The result is a quintessential architectural folly, a building of only very limited practical use – as even the besotted client eventually realized. The house proved cold in winter and subject to streaming condensation, and was unbearably hot in summer, with the mosquito-ridden site making it impossible to open the windows. All fairly predictable, but the client felt duped and – as one awaking from a dream – suddenly perceived her fairy home as an object quite absurd. Perhaps she had suddenly fallen out of love with the master. She refused to pay Mies' bill, and he had to take her to court – a most sorry end to the romance.

But their 'child' endures, and you may make of it what you will: a 'primitive hut', as eighteenth-century explorations of the origins of architecture were termed? A pavilion of delight? A monument to the minimal and the coldly rational? Whatever your view, as Edith Farnsworth discovered to her sorrow and anger, the building is hardly a practical model for modern living.

But in the end, this doesn't matter. All architects deserve such opportunities to dream in steel and glass, and Mies, when the chance came, created a toy that, in a way, not only reveals the pure perfection he had long sought but also, perhaps unexpectedly, has proved an inspiration, an architectural benchmark.

The site is no longer remote, so traffic now rumbles continuously, out of sight but not out of mind. But the building still looks crisp and pure, audacious in its almost arrogant disdain for the mundane practicalities of living. And as with all Mies buildings, the house repays close inspection. God is indeed in the details, such as the thin-as-possible steel mullions that divide the large panes of glass. Also impressive are the limited palette of forms and materials, and the treatment of inside and outside spaces in similar manner. For example, all floors, in the house or on the deck, are paved in the same white travertine. This breaking down of the boundaries between inside and outside, made clear when all the windows are open, is evidence of Mies' debt to another exemplar: the traditional Japanese house.

Does the functional failure of the Farnsworth House mean that it is fatally flawed as a work of architecture? Not at all. It is now, appropriately, a house museum, and so its function is not to provide a work-a-day home but to feed the imagination and offer the same impractical delight that can be found in the contemplation of a glittering jewel of great worth.

93

HOPKINS HOUSE
Hampstead, London | 1976 | Michael Hopkins

The term 'High Tech architecture' is somewhat nebulous, but it generally suggests an inclination to experiment with pioneering technology in the construction and servicing of buildings. But there have been few architectural epochs or movements when this was not, more or less, the case, and thus it would be logical to term early Gothic or late nineteenth-century engineered construction that resulted in early skyscrapers as 'High Tech'. Despite the vagaries of meaning, however, there are a number of key factors that can be said to define High Tech architecture in its early phase – essentially the two decades after 1967 – as it emerged in the United Kingdom, which was to all intents and purposes its breeding ground. The movement's early masters included Norman Foster, Richard Rogers, Renzo Piano, Nicholas Grimshaw, Jan Kaplicky, and Michael and Patti Hopkins. These architects generally had only informal relationships with each other, although Foster and Rogers worked together in Team 4 from 1963 to 1967 and designed what is arguably the first High Tech building – the Reliance Controls factory of 1967, in Swindon. Perhaps ironically, this was also the last building on which Rogers and Foster collaborated.

Given the informal and occasionally competitive relationship between the early High Tech architects, it is hardly surprising that the movement had no manifesto and no conferences. What linked them all was the belief – hardly novel – that architecture should express the spirit of its age, and for High Tech architects the defining spirit of the latter twentieth century was the application and expression of the potential offered by 'advanced' technology.

Exactly why High Tech rose and reached its initial peak of excellence in Britain during the 1970s is hard to explain. The most perceptive historian of the movement, Colin Davies, argues that 'there is something indefinably British about High Tech' because it is 'perhaps … a reflection of that British literal-mindedness that sees architecture not as a high-flown art or philosophy, but first and foremost as technique', as an activity that is essentially practical.[6] High Tech architecture can also be placed within the British tradition as outlined in the early nineteenth century by the neo-Gothic A W N Pugin, who wrote in 1841, in *The True Principles of Pointed or Christian Architecture*, that 'there should be no features about a building which are not necessary for convenience, construction and propriety' – a fair and succinct description of the High Tech approach.

The key factors which characterize High Tech architecture include several that are fundamental to early twentieth-century Modernism – for example, honesty in the expression of material and means of construction, of the technology used, and of the function served.

This honesty led to one of High Tech's most famed and notorious motifs, which is the exterior display of services, such as air-conditioning ducts, plumbing, and elevators. It can be argued that this was simply the result of the ruthless application of logic. Elements of a building that are purely functional but need maintenance and regular replacement were to be exposed and made easy to get at. This display of services also means that, for reasons of economy and reliability, the more complicated components of a building can be fabricated on a production line, shipped to site complete and fully tested, and then 'plugged in' during the final phases of construction.

In High Tech the rational display of services, along with primary structure, often becomes ornamental and in a sense the ultimate expression of the Puginian neo-Gothic stricture that all ornament in architecture 'should consist of enrichment of the essential construction of the building' – an axiom that in High Tech is extended to include the ornamentation of services and of function.

In addition, High Tech displays a preference for precision building components: metal and glass rather than bricks, stone, concrete, and mortar. This factor alone has made High Tech an essentially international style because components are acquired from the world market depending on cost and performance – an approach that contradicts the convention, expressed forcefully in the Arts and Crafts movement of the late nineteenth century, that buildings should be rooted to their sites through the use of local materials and building traditions. Related to this preference for international materials is the issue of context. To what degree can High Tech architecture be happily integrated into a historical setting, particularly in the urban context, that is characterized by the use of traditional and local building materials and techniques?

High Tech architecture also displays a determination to utilize industrial production, and to use industries other than construction as sources of technology, inspiration, and imagery. This led to a fundamental High Tech practice of 'technology transfer', wherein architects would explore the world of industrial design and production to find components for their buildings. This predilection can be seen as an echo of Le Corbusier's passion for the aircraft and automobile industries, which he viewed as inspirational and exemplary.

Other key High Tech concerns include flexibility, adaptability, and the renewability of part or of the whole of a building when demanded by changes of function or performance. A building is viewed, without sentiment, as just another disposable technological creation, so that when it fails or is outdated, it is transformed or replaced.

Although an architecture that rejects historicism, High Tech has its own history, and although determined by function, High Tech also has its own repertoire of inspirational – even romantic – imagery. Its history is the history of the functional tradition, including the mid-nineteenth-century iron and glass structure at the Palm House at Kew, the train shed at St Pancras station, and the pioneering technology-based designs of Buckminster Fuller, Pierre Chareau, and Johannes Brinkman, Leendert van der Vlugt, and Mart Stam's Van Nelle Factory of 1925–31 in Rotterdam. Another powerful inspiration was the early 1960s fantasy 'mega-structures' of architects and theorists such as Peter Cook and Archigram and Cedric Price, with their Plug-In City and Fun Palace projects. High Tech's 'romantic' imagery is derived from early twentieth-century projects by Russian Constructivists such as Yakov Chernikov and Alexander Vesnin, and by Italian Futurists, notably Antonio Sant'Elia, who admired technology, speed, and the industrial city – as well as violence and war.

More specifically, High Tech can be seen as the ultimate expression of early Modernist doctrine. For example, Le Corbusier stated in 1923 that a house is a machine for living in, but his early houses, such as the Villa Savoye, do not look like machines. In contrast, High Tech houses not only possess machine-like efficiency but also contrive to look like machines – for example, Michael Hopkins' own house and studio in Hampstead of 1976. So, in a simple sense High Tech architecture was the final realization of early Modern theory.

Hopkins House is an early and dazzling display of High Tech logic and constructional principles, and all the more dramatic because the steel and glass two-storey building is placed within a large, undulating, and leafy garden in a north London conservation area. But here the juxtaposition of tradition and organic nature with modern technological and orthogonal construction is not painful but one of the project's most pleasing and creative glories. One complements the other brilliantly so that, magically, what could have been an eyesore has become a much-loved urban ornament. The house utilizes familiar 'big' building technology on a miniature scale, which means that its somewhat incongruous exposed steel lattice girders and metal decking imbue the structure with some of the charm of a folly. The front and rear elevations are fully glazed with sliding glass doors – a transparency and accessibility echoing Mies' early 1950s Farnsworth House – and its interior is open-plan and flexible, with some privacy between rooms provided by venetian blinds and glazing. Inside and out the building is characterized by bold simplicity, a refinement of detail and the sustained application of a ruthless logic governing construction and modes of habitation.

But within High Tech's overt display of machine-like functionalism lurks a paradox. The ruthless exposure of structure and services – the hallmark of the movement – is often not the most logical, economic, or practical solution; thus, the movement's adherence to this leitmotif reveals that, in the end, High Tech is poetic architecture and not pure, rational engineering.

94

LLOYD'S OF LONDON
Lime Street, the City | 1978–86 | Richard Rogers

The epitome of large-scale and complex High Tech urban architecture is Lloyd's of London. Designed in the late 1970s by Richard Rogers and opened in 1986, it reveals, with crystal clarity, the richness and artistic and emotional potential – as well as the limitations – of High Tech. To an obvious degree the Lloyd's design is a child of the Centre Pompidou in Beaubourg, Paris, opened in 1977, that Rogers designed with Renzo Piano. The Pompidou, with its exposed structure, external escalator, and brightly coloured service ducts, cast Rogers before the public in the role of provocative architectural iconoclast. For most, the Pompidou was 'love at second sight'.[7] But despite its industrial form – a bit like an oil rig – and early teething problems, its sculpturally exposed circulation system and open, flexible plan worked. Incredibly, it possessed Parisian chic.

The Pompidou, however, was a public arts building, while Lloyd's was for City financial types. Perhaps symbolically, colour was abandoned for shades of grey and silver – the tones of convention and bullion. The essential structure also changed. The prime structure at Pompidou is a frame incorporating steel columns filled with water to prevent damage by fire, but at Lloyd's Rogers abandoned the High Tech preference for a steel structural frame and instead used an in-situ cast and precast concrete structural frame. Also the vigorous aesthetic of 'machine' architecture had to be manipulated in response to the historic nature of the City site. However, Constructivist and Archigram imagery was evoked to create a picturesque silhouette and heroic machine-age look, with vast external ducting and stainless-steel wall-crawling lifts. Part of the response to the demands of the historic site and conventional City tastes meant that the apparently ruthlessly mass-produced steel structural components of the frame were in fact beautifully crafted and gratifyingly ornamental in appearance.

The final paradox is that this functional design based on maximum flexibility has not proved particularly functional or flexible – Lloyd's is now thinking of moving out because the space is no longer suitable for its changed needs, and the external services, exposed to the elements, are costly to maintain. And in 2011 this apparent monument to machine-age architecture, disposable when no longer functional, was listed for preservation as a grade I historic building.

95

ROZAK HOUSE
Northern Territory, Australia | 2002 | Adrian Welke, Troppo Architects

The evolution of 'green' architecture, conceived to mitigate global warming and potentially catastrophic climate change, has become one of the most pressing issues of the early twenty-first century as our world appears to move closer towards irreversible, and man-made, environmental damage.

The architectural response to the problems of our increasingly energy-conscious and ecological age is very mixed, but one lightweight structure, poised within the remote and rocky terrain near Lake Bennett in Australia's tropical Northern Territory, is exemplary because it is not only an efficient riposte to many of our current environmental challenges but also possesses visual elegance and distinct architectural character

The term 'green architecture' is potentially all-embracing, but it essentially means architecture that is as non-polluting as possible – in the means by which its components are made and transported to site, in the construction process, and in the manner is which the building operates. This means a significant reduction in the highly environmentally damaging carbon dioxide or other 'greenhouse gases' emitted, which are usually produced by the consumption of fossil fuels such as petrol and oil for transportation, and oil, gas, or coal for electricity to operate machines and for conventional heating and air conditioning. In addition, construction materials should be sustainable, made only from renewable resources, and as far as possible the building should generate its own energy and dispose of its own waste in a sustainable and non-polluting manner. When the total energy consumed by a building is equal to the amount of renewable energy generated on site, the environmentally prized goal of being 'zero-rated' is achieved.

The architectural strategy suggested by this quest for green architecture generally leads to the realization of a building that is in harmony with the environmental potential of its site, oriented towards the sun for solar power and towards prevailing winds to benefit from natural ventilation; that uses materials which offer good insulation, to hold heat inside in winter and keep it out in summer; that utilizes light-weight construction to make the house lie as lightly as possible on the land; and that uses materials which are readily available locally and easily replaced without environmental damage. And not only should energy ideally come from solar or wind power, but waste should be used to gain additional energy or recycled in a useful and ecological manner.

The attempt to achieve such functional architecture in an artistically pleasing manner, or in a way that can allow a green building to fit comfortably into a traditional rural or urban context, has became the Holy Grail of ecological design. Often a green building functions well but also confirms the adage that a house is a machine for living. Green buildings too often appear mechanical, mechanistic, or even outlandish, with their architectural expression overwhelmed by the functional requirements of the 'green' brief, or by the desire to use cheap, available, and sustainable materials with high insulation characteristics. For example, straw bales as building blocks can do the job well, but hardly produce an architecture that fits harmoniously into a sensitive or traditional urban context. This difficulty is perhaps strange since much traditional architecture is, by its nature, highly sustainable and reflects an innate human preference for the use of local materials, and the quest for good insulation was second nature to economy-minded vernacular builders who were more than happy to keep fuel bills down. Pounded earth or *pisé* is a common vernacular walling material around the world, and nothing could be more cost-effective, sustainable, or sound when it comes to insulation. Also brick, although polluting to a degree to produce, is a robust, low-maintenance material and potentially very good for insulation, as is stone, which traditionally was quarried with minimal pollution.

The architects of the Rozak House claim they learned the 'lessons' of local Aboriginal structures and vernacular dwellings to create an architecture 'grounded' in the 'traditions of place' while 'reaching to the future in pursuit of environmental sustainability'.[8] The design, they say, is intended to connect 'the indoors with the out' and 'respond to the morning, the evening, the season, the heat, the cold, the sun [and] the rain',[9] which in this tropical region includes monsoons and cyclones. So the design of the house is shaped by the cyclical movements of the sun and wind.

The house is essentially three pavilions connected by boardwalks, and its incredibly light and skeletal structural form is achieved by the use of slender-section steels to give strength in a most elegant and minimal manner and resists fire and termites. In traditional manner, the house has sloping roofs to keep out rain and its floors are elevated to capture the movement of air. The roofs contain polycarbonate panels so light can penetrate inside; floors are made of Cyprus pine slats that allow air to circulate and moisture from the season-ally humid atmosphere to seep out; north-facing walls are clad with corrugated metal, while most others are formed with plywood panels or are glazed and fitted with jalousie blinds that have adjustable horizontal slats for regulating the entry of light and air. Heat comes from two active solar systems located on the roof – a solar hot-water collector and photovoltaic panels – while rain from the roof is collected and stored for future use. The architect explains this design strat-egy by pointing out that in the 'tropics less is more; architec-ture should be lightweight and elevated. The less you have of the building the better. It has to breathe and provide airflow to optimize the cooling capacity.'[10]

The extremes of the climate, and the inability to generate enough solar power to operate mechanical air conditioning or a range of familiar labour-saving domestic appliances, mean that the house dictates an energy-conscious lifestyle, one in tune with the cycles of the seasons and of the day. For example, life in the house tends to follow the daily passage of the sun, as revealed by the creation of 'sunrise' and 'sunset' rooms. This, argue the architects, 'heightens the experience of the site' and brings the occupants of the house closer to nature.

The house succeeds brilliantly as a demonstration of green design, and this has been recognized by the Royal Australian Institute of Architecture, which in 2002 gave the house its sustainable architecture award. It achieves 'zero-energy' and very low carbon emissions status, but equally importantly, rooted creatively within its site, it also possesses more than a touch of visual poetry.

CHAPTER 7

LOST AND FOUND

THE PHYSICAL REALITY OF a building – the materials and craft of its construction – is of great importance. But buildings also represent ideas, so it could be argued that their design – their spiritual as opposed to their material life – is of at least equal importance. The relative significance of these two aspects of architecture varies around the world. In the West, the materiality of a building is all-important – it is where its soul is perceived to reside. Its masonry bears the patina of age and bears witness to the passing of the years, and to the people that made and have, through the centuries, used the building. But in much of Asia and the East the focus of a building is not its material existence but its spiritual aspect; its essence is defined by its design, by the manner in which it is made and by its site. In China ancient buildings are constantly rebuilt – in a more or less authentic manner and to their initial design – and are regarded as sacred. In Japan, ancient Shinto shrines have in many cases been rebuilt in the traditional manner on a regular basis, a custom that ensures that the same building activity is repeated once in every generation so that building techniques and architectural forms can, along with religious ceremonies, be passed faithfully to posterity. For example, the minimal timber-built Ise Grand Shrine at Ise, Mie Prefecture, in Japan, is rebuilt every twenty years, so that it is forever new in its material manifestation but has not changed in its design, structural techniques, or essential function for nearly 2,000 years, and is venerated as an ancient and sacred building. Do great buildings live as powerfully in the imagination as in the material world? Successful schemes to rebuild once lost buildings suggest this is the case. Lost architecture, if rebuilt with truth and commitment, with integrity and authenticity – and ideally incorporating significant elements of the original building – can live again.

This chapter looks at five sites, cities, and buildings that reveal different aspects of the theme 'lost and found'. The first building – the 2,575-year-old Ishtar Gate from Babylon, Iraq – is now in the the Pergamon Museum, Berlin. Decayed and crumbled in antiquity, and buried for centuries, its scattered parts were found and reassembled in the late nineteenth and early twentieth centuries. Amazingly, this long-lost architectural wonder of the ancient world now once again inspires and astonishes through its design and technical virtuosity. Its recreation is a truly miraculous return from the grave. The Frauenkirche in Dresden and the Catherine Palace at Tzarskoye Selo, St Petersburg, were both reconstructed with great skill and commitment after catastrophic damage and near total destruction during the Second World War. The Euston Arch in London, a stupendous Greek Revival structure controversially demolished in 1961, could yet rise again following the recent discovery of at least 60 per cent of its original stones; while Hatra in Iraq tells the tragic story of an architectural wonder lost, found, and – in part – lost again.

96

ISHTAR GATE
Babylon, Iraq | c.575 BCE

The Ishtar Gate is one of the most artistically moving and technologically exciting survivors from the ancient world – although 'survivor' is perhaps the wrong word to use, for only a small portion remains in situ. Most of the gate's surviving fabric – including all of its finest parts – is now dispersed around the world, but a semblance of its original architectural power can be sensed within the Pergamon Museum in Berlin. Here part of the gate has been reconstructed, centuries after being lost and buried.

The Ishtar Gate – in fact a double gate – stood astride the Processional Way that cut, north–south straight and true, through the heart of the inner city of Babylon. A wonder of the ancient world and one of the eight gates into the urban sanctum, it marked the sacred and ceremonial threshold from the north, a point of entry of intense importance. The gate was constructed in around 575 BCE on the orders of Nebuchadnezzar II, the 'king of kings', who by tradition is thought to have sacked Solomon's Temple in Jerusalem in 587 BCE following a rebellion by his vassal subjects in Judea, to have carried the temple's rich and holy treasure to Babylon, and to have placed much of the Jewish population in bondage in Babylon, as outlined in the Biblical Book of Daniel. On each side of the Processional Way, running parallel with and to the east of the Euphrates, were temples and royal palaces and, legend states, the Hanging Gardens (perhaps constructed by Nebuchadnezzar II to remind his homesick wife Amytis of the verdant beauties of her Persian homeland), and the mighty ziggurat described in the Bible as the Tower of Babel.

I first went to Babylon in November 2002, in the last months of the regime of Saddam Hussein. Large sections of the fragmentary remains of the ancient city were being ruined by ignorant reconstruction, all part of Saddam's policy to cast himself as the new Nebuchadnezzar and build himself into the historic fabric of Iraq. But, thankfully, the Processional Way itself remained inviolate, much of its ancient roadway intact, with worn-paving stones laid in bitumen, and flanked by the lower walls of the brick buildings that once rose above it. On one side were steps leading down to a narrow court, above which once towered the mighty gate. The brick walls of the court remain high and remarkably well preserved, and they contain truly wonderful details. Spaced among the courses of kiln-fired bricks, still as hard as the day they were made, are large bas-reliefs showing the sacred beasts of Babylon: bulls or aurochs representing the god Adad, and dragons (with the legs of eagles and lions) representing the great god Marduk. The figures are stunning, beautifully and finely detailed, and although each is formed of many individual bricks, their outlines are continuous and coherent. To have achieved this perfection over 2,500 years ago seems a miracle, revealing the huge technical expertise of Babylonian brick makers and their ability to calculate and control shrinkage. To reach this standard of accuracy brick clay had to be mixed in a consistent manner, with ingredients strictly controlled, and then the bricks had to be fired to a precise and high temperature for an exact time, with the kiln kept at a constant and controlled heat. If all of this was not done, the wet bricks would have shrunk in an uncontrolled manner and the perfect composite images would have been impossible to make.

Mesopotamia is almost certainly where the manufacture of kiln-fired bricks started, perhaps as long as 6,000 years ago. Certainly, by the time the Old Testament was written, probably around 3,500 years ago, brick-making in the region was a well-established trade. The Book of Genesis, when outlining the making of the Tower of Babel, records that people said to one another: 'Go to, let us make brick, and burn them thoroughly. And they had brick for stone' (Genesis II: 3). A fine description – burned brick 'for stone' – but in fact the bricks were even harder than stone.

Above these figures still embedded in the wall once stood the gate, and similar figures were emblazoned upon it, but with a significant difference. The figures on the gate were coloured with glaze, so bull and dragon appear in lifelike manner, and the bricks forming the walls are glazed dark blue, with details picked out in yellow ochre. Magnificent reliefs of lions – also wrought of glazed, coloured bricks – lined the walls of the Processional Way. The surviving unglazed brick reliefs are now thought to be the remains of the initial construction, soon buried when the polychromatic gate was constructed above.

During the first two decades of the twentieth century the site was excavated by German archaeologists, and the glazed bricks discovered were carried back to Germany. The smaller part of the double gate was reconstructed in the Pergamon Museum (in reduced scale), where it stands in all its powerful polychromatic glory. Parts of the larger gate remain in store.

An inscription discovered on the gate confirms that it was constructed by Nebuchadnezzar II 'out of pure blue stone [seemingly the terms for stone-hard fired brick and stone were the same] ... magnificently adorned ... for all mankind to behold in awe'. Since glazed figures from the gate are now displayed in several museums in the USA, Canada, Sweden, Istanbul, Munich, Vienna, London, and Paris, Nebuchadnezzar's prediction that his great work would be viewed by 'all' with 'awe' was no hollow boast.

97

FRAUENKIRCHE
Dresden, Germany | 1736 | George Bahr

When completed in 1736 to the designs of George Bahr, the Frauenkirche, in the Neumarkt, was one of the outstanding Baroque buildings of northern Europe. It reflected perfectly the artistic aspirations and the religious conventions of the wealthy Protestant population of Dresden and, in its powerful domed design, made a significant contribution to the beauty of the city as a whole.

The dome became a favoured attribute for churches of ambitious design when Filippo Brunelleschi had from the 1420s to the 1440s built a dome over the crossing of the medieval Gothic cathedral in Florence. Domes are difficult to construct and to keep standing; by their nature they possess a substantial lateral thrust that has to be restrained. The nearer a dome comes to a perfect hemisphere, the greater the lateral thrust and the greater the need for substantial restraint. In response to this fact Brunelleschi gave his dome an ovoid form so that a large proportion of its weight could be carried vertically down – via stone ribs – to reduce the lateral thrust. Later Renaissance architects pondered how to do better. They puzzled over the great existing antique examples of large domes – the Pantheon in Rome and Hagia Sophia in Istanbul – but did not fully comprehend their structural nature or the complex roles played by their materials and methods of construction, and thus pursued an intriguing variety of approaches and solutions. These usually involved constructing multi-skinned domes in relatively lightweight materials – timber and lead – to reduce the dome's weight and lateral thrust, and setting the dome upon a solid and well-buttressed masonry drum.

Michelangelo's dome at St Peter's in Rome, designed in the late 1540s, is formed by two skins of brick, with the outer skin strengthened by stone ribs that are buttressed by columns on the drum from which the dome rises. But the dome is ovoid in shape – perhaps the result of Giacomo della Porta and Domenico Fontana's intervention in the 1580s, when they completed the dome long after Michelangelo's death – and so the lateral thrust is much reduced. In the mid-seventeenth century, with the main dome of Santa Maria della Salute in Venice, Baldassare Longhena achieved the ideal hemispherical dome, but only by making its outer skin of timber and lead to reduce weight and outward thrust. The timber ribs of the dome, concealed beneath its outer surface, rise from a timber ring beam reinforced at higher level with iron. Even with the use of a timber and ring beam, the outer thrust of the hemispherical dome was substantial enough to demand the construction of the large stone serpentine-shaped scrolls that give the dome a very distinct profile while also buttressing its drum.

In Paris, in 1680, Jules Hardouin Mansart came up with an ingenious – and influential – idea for the construction of the dome of Les Invalides. It was to be a truly Baroque and theatrical affair of three skins, two of which could be seen from within the church, while the third, of timber and lead and built off the middle masonry saucer dome, was designed for best effect when seen from afar. The idea of a three-skinned dome was developed in the 1690s by Sir Christopher Wren at St Paul's Cathedral in London, where a strong but unsightly brick cone, its base reinforced by a wrought-iron chain, supports the perfectly hemispherical outer dome of timber and lead, and is masked from internal view by the third dome.

In 1730s Europe, these cunningly constructed and theatrical domes were viewed with some suspicion. Increasingly, truth and beauty were seen as synonymous, and the quest was on to construct honest, structural domes in the manner of exemplary antique buildings like the Pantheon. And this is what Bahr achieved at the Frauenkirche, where the dome was a solid masonry construction, with no sleights of hand.

Bahr's magnificent church stood until disaster overtook it on the night of 13 February 1945, when 75 per cent of the historic core of Dresden was destroyed or seriously damaged by aerial attack and between 25,000 and 35,000 people were killed.

The church was left a gutted ruin that collapsed after a few days. It remained a heap of rubble during the intervening decades, when Dresden was part of the German Democratic

Republic, but in 1992, after the reunification of Germany, the decision was taken to rebuild it. It was intended not just to symbolize the recreation of central Dresden and of the newly united nation of Germany, but also to serve as a reminder of the horrors of war and to act as a gesture of peace and reconciliation.

The act of rebirth is amazing. Some 45 per cent of the stones used in the construction are original, salvaged from the site, so in a sense the Frauenkirche is not so much an archaeologically correct recreation as a radical repair. An act of love and commitment, the reborn church has character and soul – and once again its sublime dome beautifies the Dresden skyline.

98

CATHERINE PALACE
Pushkin, St Petersburg, Russia | 1748–56 |
Bartolemeno Rastrelli

The Catherine Palace, 25 kilometres (15 miles) to the south of the St Petersburg city centre in the royal park of Tsarskoye Selo, was built by the woman who was, in many ways, responsible for the completion and establishment of St Petersburg as a great city. Empress Elizabeth was the daughter of Peter the Great, who founded St Petersburg in 1703, and she put her heart into the realization of her father's dream. He had been determined to create a modern capital – 'a window onto Europe' – in the barren land around the Baltic that he had just, with great effort, reclaimed for Russia from the Swedes.

After Peter's death in February 1725 the capital of Russia was soon returned, from the cold and remote wastes of the Baltic, to Moscow. But Peter's successor, his niece Anna, moved the capital back to St Petersburg before she died, and Peter's daughter Elizabeth, when she came to the throne in 1741, committed herself to making the city work. During the late 1740s and 1750s nobles and courtiers built palaces in the city centre, near the queen's own rebuilt and enlarged Winter Palace. The queen's palace, and some of the best of the nobles' palaces (for example, the Stroganov), were the work of the same architect, the man who gave mid-eighteenth-century St Petersburg its distinct look – its powerful and potent architectural beauty.

Bartolemeno Rastrelli was born in Paris in 1700, the son of an Italian sculptor, and came to Russia in 1716, with his father, to work for Peter the Great. In the following years the young Rastrelli travelled extensively in France and Italy while also working for the Russian court and, thanks to his talent and training, was made court architect by Empress Anna in 1738. This was not a particularly timely appointment, for Anna soon died and the new regime of her sister Elizabeth cleared the court of many of Anna's favourites and advisers. But Rastrelli survived, no doubt because the gusto of his particular Baroque – colourful, powerfully detailed, and urbane – had a western European sophistication that Elizabeth thought appropriate as the architectural hallmark of her reign.

The great individual expression of Elizabeth's rule was to be the vast palace that Rastrelli was commissioned to build for this new and politically ambitious empress. It was to be one of the most impressive structures of its kind ever created. Elizabeth named the building after her mother – Catherine

– and made its design and construction the architectural focus of her life. This palace, intended to be the main summer residence of the royal family, was to be the Russian equivalent of Louis XIV's Versailles, the very epitome of power and imperial order. In this building, beauty was to be used as a political weapon and an expression of divine majesty.

The project got off to a tentative start in 1748, but by 1756 Elizabeth and Rastrelli had completed a masterpiece – a building that speaks with the authentic voice of its place and age. The architecture is Baroque, the perfect style for Elizabeth since it was rooted in Western classical culture, had become symbolic of power and authority, and was rhythmic and ordered yet capable of richness and invention.

The form of the palace is extraordinary. It's thin in plan, and incredibly long – 350 metres (1,150 ft) from end to end – because the interior is organized to present a formal route along which visitors would progress on a journey to the heart of the imperial court. The building, with its axes of power open to a limited few, remains a brilliant diagram in miniature of the mid-eighteenth-century Russian royal world.

Colour was crucial as a means of bringing a sense of Mediterranean exuberance and gaiety to the often snow-bound terrain of northern Russia. The walls of the palace are turquoise, with white architectural trim and various details now picked out in yellow ochre but once gilded. As Rastrelli himself explained, 'All is richly gilded … capitals, columns, pilasters, window pediments, statues, vases … in general everything is gilded up to the balustrade', and this included the timber glazing bars.

Inside, the palace is organized around a long and dominant processional route – an enfilade inspired by Versailles –

that stretches from virtually one end of the palace to the other, with all door openings aligned to create a spectacular internal vista. This was a parade of power and beauty, with Elizabeth, the empress, as its focus. The route was intended to mesmerize by means of its architectural grandeur – opening into room after room of dazzling design – and to reveal and confirm the hierarchy of the court, with only the highest in the land, and those most favoured by the empress, permitted to penetrate its deepest and most intimate parts.

This sensational world was virtually destroyed when German forces laid siege to St Petersburg (then called Leningrad) in September 1941. The palace was in the front line, occupied and gutted by the Germans (with by chance the chapel interior alone surviving, damaged but essentially intact) until they were forced to retreat in early 1944.

What we now see is a magical recreation. The loss of this building was intolerable to the Russian people so, within the fire-blackened walls, a lost world of beauty has been lovingly brought back to life, in authentic manner and to the highest standards of workmanship. National pride – and cultural identity – has been restored, and a great wrong has been put right.

99

EUSTON ARCH
London | 1836–37 | Philip Hardwick

The Euston Arch, a legendary London landmark, was demolished in 1961 after a short and sharp final campaign to save it. Completed in late 1837, the arch was the architectural centrepiece of Euston Station, which, when it opened in July 1837 for the London and Birmingham Railway, was the first mainline inter-city railway terminus built in any capital city in the world.

The Euston Arch was the first great architectural monument of the railway age. Taking the form of the largest Doric *propylaeum*, or gateway, ever built, it was one of the finest Greek Revival buildings in the world and a powerful declaration of the continuing belief in classicism at the very moment that the moralistic and intolerant Gothic Revival was starting to emerge as the new national style. On February 1837 the directors of the railway company declared that the 'grand and simple portico' marked the 'national character of the undertaking' to construct a railway between the great cities of London and Birmingham. Clearly, for the directors there was no irony in marking the arrival of the new technologically driven age of communication with an antique portico – on the contrary, for them it gave the mechanistic age of steam power the pedigree of culture and history.

Euston Station, with George and Robert Stephenson and the brilliant Charles Fox as its engineers, was not the world's first major railway terminus. Liverpool Road in Manchester had opened in September 1830 to serve the Liverpool and Manchester Railway, the world's first fully steam-powered intercity passenger service. The Stockton and Darlington Railway had opened in September 1825, but although passengers were carried, it was essentially a colliery railway that combined horse and steam power. George Stephenson was involved with both these enterprises as engineer and, with his son Robert, acted also as locomotive builder. London Bridge Station, serving the London and Greenwich Railway, had opened in December 1836, with its first phase, terminating at Spa Road, Bermondsey, having opened in February of the same year. So, although not strictly the pioneer of the railway terminus, Euston did much – due to its size, complexity of operations, prominence of site, and great ambition – to raise the railway terminus from a mere functional building to a noble and architecturally distinct building type and urban ornament.

In 1846 the construction of Euston Station was immortalized by Charles Dickens in *Dombey & Son*, in which Dickens described with shocked awe the 'great earthquake' that shook Camden Town as the tracks of the London and Birmingham Railway were driven and burrowed south through a deep cutting: 'houses were knocked down; streets broken through and stopped; deep pits and trenches dug in the ground; enormous heaps of earth and clay thrown up; buildings that were undermined and shaking, propped by great beams of wood'.

In its scale and mechanized energy, the construction of Euston Station symbolized modernity – for good or ill – and the rapid growth of the ruthless railway age. The improved links between the north and south of Britain, aiding communications and commerce, were the immediate benefits of the nation's pioneering railway system, and initially there were potentially huge profits to be made by the financial adventurers setting up the private railway companies. And the Euston Arch was the first great prodigy building for an exciting age, unprecedented in the scale and speed of technological development involved.

The initial design for Euston Station incorporated platforms with 60-metre (200-ft) long cast-iron roofs designed by engineer Charles Fox, but the elaborate booking hall and hotel, those uses that were to become key components of railway termini, only arrived later. What Euston Station did have from the start was its impressive portal of entry, its symbolic threshold to and from London – the Euston Arch. This was designed by Philip Hardwick, who was responsible for the architecture of the new terminus. Standing a little over 21 metres (70 ft) high, the Doric *propylaeum*, modelled loosely on that marking entry to the Acropolis in Athens, was a brilliant piece of up-to-date engineered construction.

It appeared to be built of load-bearing masonry in traditional manner, but was in fact a construction of brick, cast iron, wrought iron, and timber, faced with slabs of gritstone from Bramley Fall, Yorkshire. The drums forming the arch's four columns were not solid but hollow, each formed of four large slabs of fluted stone connected with iron cramps. This ingenious construction technique made the arch quicker and cheaper to build without diminishing its beauty or strength.

The demolition of the Euston Arch in 1961 did not mark its end – rather, it marked the beginning of a long fight to rebuild the arch and to reverse this act of shameful barbarism, perpetrated by irresponsible authorities in the face of public opposition. Reconstruction is possible because over 60 per cent of the arch's original stones have been discovered, in good condition, in an East London river; a significant number have been recovered, and a campaign launched to use these stones to rebuild the arch in authentic manner as part of the currently proposed redevelopment of Euston Station. When this happens a brutal wrong will have been put right and a structure of great beauty and meaning given back to London.

100

HATRA

Iraq | 1st century BCE to 3rd centuries CE

When I visited Hatra in November 2002 it was, in its way, one of the most thrilling and perfect archaeological sites in the world. I remember driving southwest from the city of Mosul, along a straight and dull desert road, and then suddenly getting my first glimpse of the spectacular stone-built ruins of the ancient city, shimmering in the sunlight, among wind-swept, scrubby sand-drifts. These fantastic surviving structures appeared almost as a mirage, but they were not – they were a dream of an antique classical city, with colonnades, porticoes, and pediments, designed with verve and invention. The ethereal image gradually took more solid and precise form as we drove nearer the heart of what, 1,900 or so years ago, had been one of the great trade and political centres in Mesopotamia and Arabia.

Hatra's ancient origins remain somewhat obscure, but it seems to have been founded in the third or second century BCE by the Hellenistic Seleucid Empire, and was taken over by the Parthians in the first century CE when it became a mighty trading city, wealthy and well fortified. Eventually Hatra formed the heart of an Arab kingdom, linked by culture, religion, and trade to other great Arab cities in the region including Baalbek, Palmyra, and Petra. Such was Hatra's success that it became a tempting target for Roman aggression and conquest. It was attacked by the emperors Trajan in 116 CE and Septimius Severus in 198 CE – but so formidable were the city's in-depth defences and so able its army that on both occasions the Romans called off their sieges and departed in disappointment.

Hatra's Arab population developed, with its sister cities, a distinct culture. An array of gods were venerated, including, at Hatra, the Grecian gods Hermes and Poseidon and the Midlle Eastern gods Shamesh, Nergal and Allat. These cities also all displayed a genius for architecture. The prototype was the classicism of Greece and Rome – the inevitable architectural model of the age – but in all these cities Roman classicism was fused creatively with indigenous forms and given a most vigorous and inventive expression. At Hatra – the only stone-built as opposed to brick-built ancient city in Mesopotamia – particularly impressive are the temple of Shamash, with its open pediment and erudite colonnades in the composite order, and the vast arched opening in the palace, anticipating the sixth-century CE arch at Ctesiphon and later iwans in Islamic mosques. Hatra is a stunning fusion of stupendous Hellenistic and Roman classicism with Assyrian and Parthian architecture. The results, although similar to the architecture of Hatra's sister cities, are distinct and in many ways unique. This is partly because of the wonderful images that abounded – for example, dromedaries carved into a palace wall reveal how important these animals must have been to the city's economy, while from another wall peered the head of what appeared to be a green man, his face framed by foliage. All this shows just how inventive and culturally diverse the Arab world was before the rise of Islam.

Hatra under Parthian rule became, like Palmyra, one of the great urban ornaments of the world – but also like Palmyra, it suffered a dramatic and calamitous reversal.

Palmyra, after rebelling against Roman domination, was destroyed by Rome in 273 CE. Just over thirty years earlier, in 241 CE, Hatra had been destroyed, and left a desolate ruin, by the Sassanid king Shapur I. The ruined city endured seventeen centuries of utter neglect, but in the mid-twentieth century archaeologists started to reassemble some of its main architectural wonders and, in a way, brought the lost and long-forgotten city back to life.

After my 2002 visit, I co-authored a book about my travels among the cultural treasures of the Middle East and Central Asia during a time of war and conflict. Of Hatra, I wrote that even this magnificent and remote ancient city 'could not be immune to the perils of war and its aftermath' and hoped that the ruined condition of some of its buildings 'would not prove prophetic for this still-proud, shimmering city'.

Sadly, my fears were well grounded. Hatra survived the Gulf War of 1991, the US-led Invasion of Iraq of 2003, and a subsequent decade of chaos and looting in the region. But it has not survived unscathed the rise of the fanatical, vengeful, and intolerant organization called Islamic State (IS), which sought to make the north of Iraq part of a Sunni Muslim caliphate and, to achieve this goal, made terror and brutal violence – usually inflicted on the helpless, be they ruined cities or captives – the core policies of its regime. The full extent of the damage inflicted by IS on the noble ruins of Hatra in 2015 is still being catalogued, but broadly it destroyed or looted virtually all the city's rich array of sculpture and engraved imagery – including the head of the Green Man – but the walls and buildings, although threatened with being bulldozed, appears to have largely survived. This is a blessing but the loss of the art has done much to rob the ruins of meaning and beauty.

It is possible to place IS's attack on history and on urban beauty – on civilization – in context. Reflecting on past acts of barbarism it is clear that, as with IS, such destruction has often been the result of religious intolerance or a thirst for revenge, for ideological reasons, to punish, to evoke terror, for political gain, and for plunder. Many reports suggest that IS has long been looting and selling the more portable and valuable antiquities of the area it controls. In addition, such destruction has historically been aimed at gaining military advantage or at utterly eradicating a once-feared foe. The Romans destroyed Carthage in the mid-second century BCE as an act of revenge and to ensure the obliteration of a rival power; the Assyrian king Sennacherib levelled Babylon in 690 BCE as punishment for a revolt, and the Romans slighted Palmyra in 273 CE; Alexander the Great destroyed Persepolis in 330 BCE, probably in retaliation for Persian outrages in Greece and to eradicate Persia as a world power; Genghis Khan and his Mongol 'horde' laid waste in the early thirteenth century to thriving cities in Central Asia and China to sow the seeds of terror and to establish the Mongol nomadic empire as the great power in the region. In the Spanish Civil War and the Second World War, cities were in the front line and were often targeted for destruction for a wide range of reasons – to instil fear, to cripple industry, to weaken an enemy's aggressive or defensive abilities, to sap morale and destroy the will to resist. The Fascist aerial attack on Guernica in Spain in April 1937 shocked the world because of its indiscriminate nature and clear aim to create terror. This policy echoed the indiscriminate Japanese bombing of Chinese cities in Manchuria from 1931 and was continued in Nazi attacks in 1940 on such cities as Rotterdam in The Netherlands. Partly in response, the British, and then to a lesser degree the Americans, evolved their own policy to level German cities. The stated aim was to destroy enemy war industries and communication networks through strategic bombing offensives against cities, but since precision bombing by daylight was very difficult and dangerous, the British turned to the carpet- or area-bombing of entire city districts by night. Gradually, in the atmosphere of an ever-escalating total war of survival against an evil regime, the destruction of enemy cities and of their communities of workers became increasingly an end in itself, with the ultimate aim being the annihilation of the enemy's ability – physically or mentally – to continue the war, a policy that found its ultimate expression in the destruction in August 1945 of Hiroshima and Nagasaki, Japan. The military effectiveness and ethics of the allied strategic bombing campaign remain debated, but what's certain is that, in the prosecution of this policy, some of the most world's most beautiful historic cities, such as Dresden and Lübeck, were badly damaged or destroyed.

Arguably the most appalling single example of city destruction during the Second World War was Hitler's calculated obliteration of Warsaw after the August 1944 uprising in which over 200,000 Polish fighters and civilians died. The aim was to punish the Poles for their resistance and to destroy their culture and identity by levelling their capital city.

The IS attacks in 2015 on the ancient cities of Nineveh, Nimrud, and Hatra in northern Iraq and then on Palmyra in Syria – culturally rich but uninhabited archaeological sites – are more complex and harder to define. At one level they represent an attack by fundamentalist Salafist/Wahhabist Sunni Muslims on images of 'living beings' that, no matter how ancient and mythic (such as the winged bull deities at Nineveh and Nimrud), could be seen as 'idols'. As such, a hard-line fundamentalist Muslim could perceive these works of ancient art as contradictions of the Koranic edicts – inspired by the second of the Ten Commandments – that man must not create nor worship nor even be distracted by 'graven images'. It was this same conviction that, in sixteenth- and seventeenth-century Europe, inflamed Puritans to destroy Roman Catholic religious imagery that they saw as idolatrous and 'superstitious'. But these IS attacks are more than yet another outbreak of iconoclasm. They have also, as with the thirteenth-century Mongols, involved the destruc- tion of architecture as well as of images, suggesting a deep and dark determination to eradicate memory and history and all evidence of life, culture, and religion before the coming of the Islamic State. IS, in the execution of its cultural terrorism and in its murderous methods and the utter intolerance it displays to people, places, and architecture that do not conform to its perceptions of righteousness, is in many ways unprecedented. Hatra, a city created before Islam, had until March 2015 survived unmolested for nearly 1,400 years within the Muslim world. No Islamic power before had argued that its ancient and beautiful sculpture was idolatrous or its archi- tecture an affront. The brutish, base, essentially stupid, and utterly unnecessary assault upon the vulnerable, defenceless, and sublime remains of Hatra is a sobering reminder that nothing – certainly no architecture or art, no matter how beautiful or ancient – is safe in this increasingly volatile and hostile world.

GLOSSARY

Abutment The end supports of a structure that carry weight to the ground. Also called buttresses. In bridges, the supports that help carry the load from the carriageway.

Apse A niche or recess, often curved in plan, set in a wall.

Architrave Part of a classical entablature that is in essence and origin a structural beam ornamented in a manner to throw off rainwater. The upper part of the entablature is a cornice, invariably topped by a projecting serpentine moulding designed to throw rainwater away from the wall, then a frieze, and at the bottom, the architrave. Also the name of the mouldings – derived from entablature mouldings – framing a door or window.

Buttress *See* **Abutment.**

Capital The topmost part of a column, often ornamental and indicative of the classical orders.

Colonnade A row of columns supporting a flat entablature or arches.

Corbel A projection from a wall to help support a structure above, such as roof timbers.

Cornice The top part of a classical entablature (see **Architrave**). Often used in isolation to ornament the top of an external elevation, usually at the junction with the roof, or the top of an internal wall at the junction with the ceiling.

Crepidoma In classical architecture, the platform on which the superstructure, usually a temple, sits. The platform generally consists of three levels, each decreasing in area as they rise, to form three steps.

Cyma Classical moulding of sinuous concave and convex profile.

Dentil A detail, reminiscent of teeth, in a classical cornice.

Enfilade A set of rooms in which doors are aligned to create a vista, typical of seventeenth- and eighteenth-century Baroque architecture.

Entablature *See* **Architrave.**

Finial An ornament on the top of a building. In classical architecture, this is typically placed above the entablature, parapet wall, or balustrade. Often in the form of an urn or pineapple.

Flute A concave vertical recess in the shaft of a column.

Glacis In military architecture, a steeply sloping surface in front of a defensive wall designed to prevent attackers from approaching near the wall, tunnelling under it, or using battering rams or scaling ladders.

Gutta (*pl.* **guttae**) Pyramidal or conical peg-like details set below triglyphs in a frieze of the classical Doric order. Thought to be derived from timber pegs securing beams in place in timber-built architecture that formed the prototypes for stone-built Grecian architecture.

Lintel A horizontal support – in any suitable material – across the opening of a door or window.

Loggia A room with at least one side open and usually defined by a colonnade, piers, or an arcade of Gothic or classical design.

Machicolations In military architecture, projections from a wall, usually supported by corbels, that include horizontal openings through which defenders can assault attackers below.

Mullion A vertical division – of timber or masonry – in a window opening, to which hinged casements of frames containing panes of glass are fixed.

Naos The inner and most sacred area of a Greek or Roman temple, usually where a cult figure was located. Also called the cella.

Narthex A forecourt, antechamber, porch, or room at the west – or entrance – end of a Christian church, forming the start of the route to the altar.

Nave The main, central, and usually highest space in a building of basilica form, usually flanked by aisles and separated from them by screens of columns, piers, or shafts.

Oculus A circular and eye-like window and, more particularly, the circular opening at the crown of a dome.

Orders In classical architecture, the five slightly different styles of design, usually each appropriate for a different function. The orders are most easily distinguished by the design of their columns. The three original Greek orders are the Doric, Ionic and Corinthian. The Romans added the Tuscan and Composite orders.

Oriel A projecting structure, rising from the ground but more usually canti-levered or supported by corbels containing windows and contrived to form a notable architectural feature.

Parterre In gardening, a level space occupied by an ornamental, and usually geometric and symmetrical, arrangement of flowerbeds.

Pediment A triangular or curved form set above an entablature – often supported by columns or pilasters – marking the main entrance to, or forming the central feature of, a classical building. The form is derived from the functional expression of the triangular end of a pitched roof.

Pendentive Arches or vaults, of triangular and concave form, that allow a dome to rise off a lower structure of square plan and usually cubical form.

Piano nobile The main floor of a building, usually a house. Derived from Renaissance practice in which, typically, the first floor was treated as the piano nobile and had the greatest floor-to-ceiling height, the richest interior decoration, and the most ornamental exterior decoration, especially on its windows, which would be the highest in the elevation.

Pilaster A flat equivalent to a column, usually in classical architecture furnished with a capital in one of the orders, like a column. Structurally, a buttress, strengthening the wall to which it is attached and helping to carry a load imposed on that wall. Sometimes used without capitals, as a means of strengthening and articulating a wall.

Piloti A column, often made of steel-reinforced concrete, used without orna-mentation in twentieth-century Modernist architecture.

Portico A sheltered entrance or porch leading to a building or a covered way. In classical architecture, often formed by the projection of a pediment or entablature supported on columns.

Propylaeum In Greek architecture, a noble gate, generally leading to a sacred or most important precinct. Usually formed with columns and piers supporting a pediment or entablature.

Purlin Part of a roof structure – one of the timbers, running parallel to the top of the wall and to the roof ridge, helping to support the rafters.

Pylon In Egyptian architecture, a monumental gate formed by a pair of towers with sloping sides, like flat-topped and truncated pyramids.

Quatrefoil In Gothic architecture, a motif formed with four cusps or lobes – typically a window or opening framed by tracery and incorporating four cusps, sometimes in the simple form of four circles, partly overlap-ping to create a four-leaf clover motif.

Spandrel The area between the outer curves of arches, or between arches and a ceiling or vault.

Squinch A structure – straight, curved, or arched – designed to reconcile a square-plan base with a crowning dome. See **Pendentive**.

Stanchion A vertical support, usually of very simple form and of cast iron, supporting roofs or superstructure generally.

Stoa A roofed colonnade.

Stupa In Buddhist architecture, a mound or dome, usually of conical form, evoking the image of the sacred Mount Meru.

Stylobate The continuous base supporting the rows of columns framing or fronting a classical temple. Sometimes placed on, or part of, the crepidoma.

Transept Projections from, and set at right angles to, the nave and aisles of a Christian church of basilica plan. The transepts are placed towards the altar end of the church so that, in plan, the church has the form of the Christian Latin cross.

Triforium A high-level passageway or gallery that is placed above the aisles, and runs parallel to the nave and sometimes around the transepts and chancel, of a Christian church of basilica plan. Used for religious ritual, including stalls for choirs. Also found in some pre-Christian basilicas.

Volute In classical architecture, a spiral scroll, perhaps of ram's horn appearance, that is the principal feature of the Ionic capital. Also used, at a smaller scale and in a more subservient role, in the capitals of the Corinthian and Composite orders.

REFERENCES

Chapter 1

1. Margaret Murray, *Egyptian Temples*, 1931, p. 25
2. *Vitruvius: The Ten Books of Architecture*, ed. Morris Hicky Morgan, 1960 edition, pp. 104, 106
3. *Ibid*, pp. 103–04
4. *I Quattro Libri dell'Architettura*, trans. Isaac Ware, 1738, p. 41
5. *The Builder*, 16 June 1866
6. *The Buildings of England: South Lancashire*, 1969, p. 177
7. 'The Tyranny of the Skyscraper' chapter in *Modern Architecture*, Princeton University Press, 1931, p. 85
8. Cited in Leland M. Roth, *Understanding Architecture: Elements, History and Meaning*, 1993
9. *Bauhaus*, 2000, ed. Fiedler and Feierabend, p. 17
10. Quoted by Kenneth Murchison, an architect colleague of Van Alen – see Neal Bascomb, *Higher: A Historic Race to the Sky and the Making of a City*, Random House, New York, 2003, p. 45
11. Frederick Etchells' English translation, *Towards a New Architecture*, 1927, p. 205
12. David P. Billington, *Maillart's Bridges*, pp. 136–37

Chapter 2

1. *Vitruvius*, ed. Morris Hicky Morgan, 1914, p. 26
2. See T E Lawrence, *Crusader Castles*, ed. Denys Pringle, Oxford, 1988
3. Hugh Kennedy, *Crusader Castles*, Cambridge University Press, 1994, p. 154
4. *Ibid*, pp. 145–47
5. Francis Yates, *The Occult Philosophy in the Elizabethan Age*, 1979, p. 94 in 1999 edition
6. Kanto Shigemori, *Zashiki: The Japanese House*, Kyoto, 1988, p. 9

7. *Country Life*, 15 October 1970

8. Alan Hess, *Frank Lloyd Wright: The Houses*, Rizzoli, New York, 2005, p. 13

9. Lecture at Pratt University, Brooklyn, New York, autumn 1973, cited in *Louis I. Kahn: Writings, Lectures, Interviews*, ed. Alessandra Latour, Rizzoli, New York, 1991, p. 326

10. Joseph Rykwert, *Louis Kahn*, Abrams, New York, 2001, p. 158

11. Talk given in Milan in January 1967, cited in *Louis I. Kahn: Writings, Lectures, Interviews*, ed. Alessandra Latour, Rizzoli, New York, 1991, p. 226

12. *Ernö Goldfinger: Works 1*, compiled by James Dunnett and Gavin Stamp, Architectural Association, London, 1983, p. 7

Chapter 3

1. See Douglas Knoop and G P Jones, *The Genesis of Freemasonry*, 1949

2. See 1 Kings 14: 25–26 and 2 Chronicles 12: 2–9, in both of which he is named Shishak

3. William Fitzstephen, *Descriptio Londoniae*, 1174–83

4. *On the Art of Building in Ten Books*, Book 7, ed. Joseph Rykwert, MIT Press, 1988, p. 196

5. Quoted in Kenneth Frampton, *Modern Architecture: A Critical History*, Thames & Hudson, London, 1985, p. 100

6. *Mystical Themes in Le Corbusier's Architecture in the Chapel Notre-Dame-du-Haut at Ronchamp*, Mellen Studies in Architecture, Ontario, 2000, p. 11

7. *Ibid*, p. 11

8. *Ibid*, p. 93

9. *Ibid*, p. 11

Chapter 4

1. See Mary Ellen Miller, *Maya Art and Architecture*, Thames & Hudson, London, 1999, pp. 40–41

2. 1765 edition, p. 119

3. Tobias Smollett, *The Expedition of Humphrey Clinker*, 1771, edited and introduction by Lewis M. Knapp. Oxford University Press, Oxford, 1984 edition, pp. 34–35

4. Tim Mowl and Brian Earnshaw, *John Wood: Architect of Obsession*, Millstream Books, Bath, 1988, pp. 198–202

Chapter 6

1. *Amerikaansche Reisherinneringen*, 1912, p. 157, quoted in *Hendrik Petrus Berlage: Complete Works*, ed. Sergio Polano, Butterworth, 1988, p. 19

2. *Ibid*, p. 19

3. August Sarnitz, *Wagner*, Taschen, 2005, p. 7

4. Quoted in Nikolaus Pevsner, *Pioneers of The Modern Movement*, 1936, pp. 174–5 of 1977 edition, retitled *Pioneers of Modern Design*, Penguin Books, London

5. Matthew Teitelbaum et al., *Frank Gehry in Toronto: Transforming the Art Gallery of Ontario*, Merrell, New York, 2009, p. 11

6. Colin Davies, *High Tech Architecture*, Rizzoli, New York, 1988, p. 6

7. *National Geographic*, October 1980, p. 469

8. Mary Guzowski, *Towards Zero Energy Architecture: New Solar Design*, Laurence King Publishing, London, 2010, p. 129

9. *Ibid*, p. 315

10. Quoted in Guzowski 2010, p. 131

Chapter 7

1. Dan Cruickshank and David Vincent, *People, Places and Treasure Under Fire*, BBC Books, London, 2003, p. 94

BIBLIOGRAPHY

The relevant building number is provided in brackets at the start of each reference.

[23] Daniel M. Abrahamson, *Skyscraper Rivals*, Princeton Architectural Press, New York, 2001.

[21] Xenia Adjoubei and Robert Hill, 'Ruins of Utopia', *The Architectural Review*, Vol. 233, No. 1394, April 2013, pp. 88–93.

[65] Isabel Artigas, Antoni *Gaudí: Complete Works*, Taschen, Cologne, 2007.

[54] Warwick Ball and Leonard Harrow (eds), *Cairo to Kabul: Afghan and Islamic Studies. Presented to Ralph Pinder-Wilson*, Melisende, London, 2002.

[23] Neal Bascomb, *Higher: A Historic Race to the Sky and the Making of a City*, Random House, New York, 2003.

[25] David P. Billington, *Maillart's Bridges: The Art of Engineering*, Princeton University Press, New York, 1979.

[73] Iain Browning, *Palmyra*, Chatto & Windus, London, 1979.

[98] William C. Blumfield, *A History of Russian Architecture*, Cambridge University Press, Cambridge, 1993.

[27] Robert Browning, 'Andrea del Sarto, Called the Faultless Painter', 1855.

[51] Tons Brunes, *The Secrets of Ancient Geometry and Its Use*, Rhodos, Copenhagen, 1967.

[21] Victor Buchill, 'Moisei Ginzburg's Narkomfin Communal House in Moscow: Contesting the Social and Material World', *Journal of the Society of Architectural Historians*, Vol. 57, No. 2, June 1998, pp. 160–81.

[73] Ross Burns, *Monuments of Syria: An Historical Guide*, I.B. Tauris, London, 1994 edition.

[58] James W. P. Campbell, *The Library: A World History*, Thames & Hudson, London, 2013.

[24] Le Corbusier, *Vers Une Architecture*, Editions Cres, Paris, 1923; English edition, *Towards a New Architecture*, translated by Frederick Etchells, 1927.

[68] M A Couturier, M R Capellades, L B Rayssiguier, and A M Cocagnac, *Les Chapelles du Rosaire a Vence Par Matisse et de Notre-Dame-Du-Haut a Ronchamp Par le Corbusier*, Paris, 1955. Dan Cruickshank (ed.), *Sir*

Banister Fletcher's A History of Architecture, 20th edition, Architectural Press, Oxford, 1996.

[52] Dan Cruickshank, *The Story of Britain's Best Buildings*, BBC Worldwide, London, 2002.

[100] Dan Cruickshank and David Vincent, *People, Places and Treasure Under Fire: Afghanistan, Iraq and Israel*, BBC Books, 2003.
Dan Cruickshank, *Around the World in 80 Treasures*, Weidenfeld & Nicolson, London, 2005.
Dan Cruickshank, *Adventures in Architecture*, Weidenfeld & Nicolson, London, 2008.

[25] Dan Cruickshank, *Bridges: Heroic Designs that Changed the World*, Collins, London, 2010.

[33] Dan Cruickshank, *The Country House Revealed*, BBC Books, London, 2011.
William J R Curtis, *Modern Architecture since 1900*, Phaidon, Oxford, 1996 edition.

[94] Colin Davies, *High Tech Architecture*, Rizzoli, New York, 1988.

[26] Colin Davies, *Key Houses of the Twentieth Century*, Laurence King Publishing, London, 2006, pp. 66–7.

[90] Peter Davey, *Arts and Crafts Architecture*, Architectural Press, London, 1980, Phaidon 1997 edition.

[43] James Dunnett and Gavin Stamp, *Ernö Goldfinger: Works 1*, Architectural Association, London, 1983.

[54] Louis Dupree, *Afghanistan*, Princeton University Press, Princeton, New Jersey, 1973.

[54] Nancy Hatch Dupree, *An Historical Guide to Afghanistan*, Afghan Tourist Organization, 1977 (second edition).

[21] Aurora Fernandez Per and Javier Mozas, *Ten Stories of Collective Housing*, A+T Architecture Publishers, Spain, 2013, pp. 86–111.

[19] Jeannine Fiedler and Peter Feierabend (eds), *Bauhaus*, Konemann, 2000.

[19] Jeannine Fiedler, *Bauhaus*, H F Ullmann, 2006.

[54] Finbarr Barry Flood, *Objects of Translation: Material Culture and Medieval 'Hindu-Muslim' Encounter*, Princeton University Press, 2009.

[54] Finbarr Barry Flood, 'Ghurid Monuments and Muslim Identities: Epigraphy and Exegesis in Twelfth-Century Afghanistan', *Indian Economic and Social History Review*, Vol. 42, No. 3, 2005, p. 273.

[17] Kenneth Frampton, *Modern Architecture: A Critical History*, Thames & Hudson, London, 1980.

[98] Arthur and Elena George, *St Petersburg: A History*, Sutton, Stroud, 2004.

[27] Paul Goldberger, *The City Observed: New York; a Guide to the Architecture of Manhattan*, Random House, New York, 1979.

[95] Mary Guzowski. *Towards Zero Energy Architecture: New Solar Design*, Laurence King publishing, London, 2010.

[73] Dr William Halifax, *Report on a Voyage to Palmyra*, Philosophical

Transactions of the Royal Society, London, 1695.

[54] Stanley Ira Hallet and Rafi Samizay, *Traditional Architecture of Afghanistan*, Garland STPM Press, New York and London, 1980.

[27] Carol Herselle Krinsky, 'The Skyscraper Ensemble in Its Urban Context', in Roberta Moudry (ed.), *The American Skyscraper: Cultural Histories*, Cambridge University Press, Cambridge, 2005.

[39] Alan Hess, *Frank Lloyd Wright: The Houses*, Rizzoli, New York, 2005.

[54] Robert Hillenbrand, *Islamic Art and Architecture*, Thames & Hudson, London 1999.

[54] Robert Hillenbrand, *Islamic Architecture: Form, Function and Meaning*, Columbia University Press, New York, 1994.

[5] Paul Holberton, *Palladio's Villas: Life in the Renaissance Countryside*, John Murray, London, 1990.

[29] Hugh Kennedy, *Crusader Castles*, Cambridge University Press, 1994.

[84] Malcolm Kirk, *The Barn: Silent Spaces*, Thames & Hudson, London, 1994.

[14] Peter Gwillim Kreitler, *Flatiron*, American Institute of Architects Press, Washington D.C., 1990.

[54] Edgar Knobloch, *Monuments of Central Asia*, I B Tauris, London, 2001.

[68] Isabel Kuhl, *50 Buildings You Should Know*, Prestel, Munich, 2007.

[42] Alessandra Latour, ed. *Louis I. Kahn: Writings, Lectures, Interviews*, Rizzoli, New York, 1991.

[29] T E Lawrence, *Crusader Castles*, new edition with introduction and notes by Denys Pringle, Clarendon Press, Oxford, 1988.

[98] W. Bruce Lincoln, *The Romanovs: Autocrats of all the Russias*, Weidenfeld & Nicolson, London, 1981.

[98] Philip Longworth, *The Three Empresses: Catherine I, Anne and Elizabeth of Russia*, Constable, London, 1972.

[17] Jules Lubbock, 'Adolf Loos and the English Dandy', *Architectural Review*, August 1983, pp. 43–9.

[4] Charles R Mack, *Dictionary of Architecture*, St James's Press International, 1993, pp. 549–51.

[67] Ivan Margolius, *Church of the Sacred Heart: Joze Plecnik*, Phaidon, Oxford, 1995.

[54] André Maricq and Gaston Wiet, *Le Minaret de Djam: La Découverte de la Capitale des Sultans Ghorides (XIIe-XIIIe Siècles)*, Mémoires de la Délégation Archangélique Française en Afghanistan, XVI, Paris, 1959.

[73] Henry Maundrell, *A Compendium of a Journey from Aleppo to Jerusalem at AD 1697 by Henry Maundrell*, Oxford, 1703.

[74] Mary Ellen Miller, *Maya Art and Architecture*, Thames & Hudson, London, 1999.

[77] Tim Mowl and Brian Earnshaw, *John Wood: Architect of Obsession*, Millstream Books, Bath, 1988.

[1] Margaret A. Murray, *Egyptian Temples*, Sampson Low, Marston & Co, London, 1931.

[78] Peter Murray, *The Saga of Sydney Opera House*, Spon Press, London, 2004.

[8] Francesco Nevola, *Soane's Favourite Subject: The Story of the Dulwich Picture Gallery*, Dulwich Picture Gallery, 2000.

[96] Joan Oates, *Babylon*, Thames & Hudson, London, revised edition 1979, reprint with updated bibliography 2008.

[5] Andrea Palladio, *I Quattro Libri dell'Architettura,1570*; English edition Isaac Ware, *The Four Books of Architecture*, London, 1738, Dover edition with introduction by Adolf K. Placzek, New York, 1965.

[26] Juhani Pallasmaa with Andrei Gozak, *The Melnikov House*, trans. Catherine Cooke, Academy Editions, London, 1996.

[89] Nikolaus Pevsner, *Pioneers of the Modern Movement from William Morris to Walter Gropius*, Faber & Faber, London, 1936; republished as *Pioneers of Modern Design*, Museum of Modern Art, New York, 1949, 1977 edition, Penguin Books, London.

[54] Ralph Pinder-Wilson, *Studies in Islamic Art*, Pindar Press, London, 1985, pp. 89–102, Chapter 7, 'The Minaret of Mas'ud III at Ghazni'.

[89] V. Horvat Pintaric, *Vienna 1900: The Architecture of Otto Wagner*, Bestseller Publications, London, 1989.

Sergio Polano (ed.), *Hendrik Petrus Berlage: Complete Works*, Butterworth, Oxford, 1988.

[26] Grant Prescott, 'Melnikov House', *Journal of Architectural Conservation*, Vol. 19, No. 1, March 2013, pp. 71–4.

A. W. N. Pugin, *The True Principles of Pointed or Christian Architecture*, 1841.

Leland M Roth, *Understanding Architecture: Its Elements, History and Meaning*, Icon Editions, New York, 1992.

[42] Joseph Rykwert, *Louis Kahn*, Abrams, New York, 2001.

[4] Joseph Rykwert (ed.), *Leon Battista Alberti: On the Art of Building in Ten Books*, MIT Press, 1988.

[89] August Sarnitz, *Wagner*, Taschen, Cologne, 2005.

[24] Jacques Sbriglio, *Le Corbusier: The Villa Savoye*, Foundation Le Corbusier, Paris, 1999.

[73] Dr Thomas Shaw, F.R.S. *Travels in Barbary and the Levant*, Oxford, 1738, revised edition 1757.

[27] Roger Shepherd (ed.), *Skyscraper: The Search for an American Style, 1891–1941*, McGraw-Hill, New York, 2003.

[88] Pieter Singelenberg, *H. P. Berlage, Idea and Style: The Quest for Modern Architecture*, Haentjens, Dekker & Gumbert, Utrecht, 1972.

[54] Janine Sourdel-Thomine, *Le Minaret Ghouride de Jam*, Diffusion de Bocard, Paris, 2004.

[2] Tony Spawforth, *The Complete Greek Temples*, Thames & Hudson, London, 2006.

[54] Freya Stark, *The Minaret of Djam: An Excursion in Afghanistan*, John Murray, London, 1970.

[91] Matthew Teitelbaum et al., *Frank Gehry in Toronto: Transforming the Art Gallery of*

Ontario, Merrell, New York, 2009.

[7] Barrie Trinder, *The Industrial Revolution in Shropshire*, 3rd edition, Chichester, Phillimore, 2000.

[22] Dominique Vellay, *Le Maison de Verre: Pierre Chareau's Modernist Masterwork*, Thames & Hudson, London, 2007.
Vitruvius, *The Ten Books of Architecture*, ed. Morris Hicky Morgan, 1960 edition, pp.104, 106.

[78] Anne Watson (ed.), *Building a Masterpiece: The Sydney Opera House*, Lund Humphries, London, 2006.
Richard Weston, *The House in the 20th century*, Laurence King, London, 2002.

[22] Richard Weston, *Key Buildings of the 20th Century*, 2nd edition, Laurence King Publishing, London, 2010.
Richard Weston, *Modernism*, Phaidon, London, 1996.

[78] Richard Weston, *Utzon: Inspiration, Vision, Architecture*, Hellerup, Edition Blondel, 2002.

[2] Mark Wilson Jones, *Origins of Classical Architecture: Temples, Orders and Gifts to the Gods in Ancient Greece*, Yale University Press, New Haven, 2014.

[77] John Wood, *An Essay Towards a Description of the City of Bath*, 1765 [1742], Kingsmead Reprints, Bath, 1969.
Frank Lloyd Wright, *Modern Architecture*, Princeton University Press, 1931.

[31] Francis Yates, *The Occult Philosophy in the Elizabethan Age*, Routledge Classics, London, 1979, 1999 edition.

INDEX

A

Aalto, Alvar 157–60, 287
Abbas, Shah 231–4
Abd al-Malik ibn Marwan,
  Caliph 185
Abraham 186, 205
Acropolis (Athens) 341, 369
Adler, Dankmar 14, 61–63,
  65
AEG Turbine Hall (Berlin)
  72–4
Afghanistan 168, 192, 203–4
Agra (India) 168, 234–7
Agrippa, Heinrich Cornelius
  125–6
Ahlers-Hesterman, Friedrich
  332
Al Qaeda 321
al-Andalus 120–1
Albert, Prince 46
Alberti, Leon Battista 27–32,
  222
Albi cathedral 163
Alhambra (Granada) 112,
  120–3
Allen, Ralph 281–2
Alleyn, Edward 43
Altes Museum (Berlin)
  138–40, 148, 157
Amitrano, Adolfo 154–5
Amsterdam, Beurs van
  Berlage 310, 326–9
Amun-Re 172
Amytis, Queen 358
Andrew II, King of Hungary
  117
Andronicus of Cyrrhus 114
Angkor Thom (Cambodia)
  168, 206–9

Angkor Wat (Cambodia) 207
Anna, Empress 364
Anthemius of Tralles 180–1
Anubis 172
Apadana (Persepolis) 259
apartment blocks
  Moscow 91, 104,
  Vienna 310, 332
Archigram 347, 349
architecture
  choice of materials 309
  definition of 1
  pioneer 13
  rebuilding once-lost 355
  rhetoric 111
  sacred 167
  size 291
  urban 251
Architecture Parlante 111, 113
Aretas IV, King of the
  Nabateans 262
Ark of the Covenant 173–4,
  298
Arkadia 22
Art Deco 97, 105
Art Gallery of Ontario
  (Toronto) 310, 336–40
Art Nouveau 62, 72, 75, 169,
  238, 332–3
Arts and Crafts movement
  153, 334, 346
Arup, Ove 287–8
Asplund, Erik Gunnar
  147–9, 287
Assyrians 259, 270, 302,
  371, 373
Athens 21–2,113–6, 159,
  341, 369
Atwood, Charles 54, 61, 65

Aurelian, Emperor 268
Australia
  Rozak House (Northern
    Territory) 350–4, 311
  Sydney Opera House 253,
    286–9
Austria
  apartment blocks (Vienna)
    330–333
  Steiner House (Vienna)
    74–7

B

Baalbek (Lebanon) 370
Babylon 5, 174–175, 255, 261
  Ishtar Gate 357–60
Badger, Daniel D 50
Bage, Charles 14, 39–41
Bahr, George 360, 362
Baird, John 49
Baker, Bill 293
Balfron Tower (Poplar,
  London) 164, 166
Bangladesh 10, 160–3
Bank of England 44
Bank of Manhattan Building
  (New York) 95–96
Barbaro, Daniele 35
Barcelona 169, 237
Bardiya 257
Barley Barn (Cressing
  Temple, Essex) 311–5
Barlow, Henry 55–60
barns 311–4
Baroque architecture 100,
  131–4, 143, 244, 334,
  360–5
Barrier de la Villette (Paris)
  149

Barry, Charles 58
Basilica of San Lorenzo
(Florence) 218–20
New Sacristy 226–30
Basilica of Septimus Severus
(Leptis Magna) 264–6
Bassae 13, 19–24
Bath 252, 280–6
'Battle of the Styles' 60
The Bauhaus (Dessau) 81–5
Bauhaus school 72, 81–5,
105
Beaux Arts classical style
336–7, 339
Bedford, Francis Russell, 4th
Earl of 280–1
Behrens, Peter 72–3
Beijing 291, 299, 304–7
Berlage, Hendrik Petrus
310, 326–9
Berlin
AEG Turbine Hall 72–4
Altes Museum 138–40,
148, 157
Bernini, Gian Lorenzo 37
Beton Brut 9
Beurs van Berlage
(Amsterdam) 326–9
Bhutan 6, 135–7
Bibliothèque National de
France (Paris) 113
Bibliothèque Sainte-
Genèvieve (Paris) 48
Bijvoet, Bernard 91
Billingsley, Sir Henry 124
Billington, David 101
BindesbØll, Michael
Gottlieb 141
Bingham, Hiram 275
Bladud, King of Bath 283–4
Bloemenwerf House
(Uccle) 75
Bocabella i Verdaguer, Josep
Maria 238
Bogardus, James 49
Bon, Giovanni and
Bartolomeo 315–6
Bonaparte, Napoleon 57, 155
Borromini, Francesco 37, 244
Borsippa (Iraq) 5, 255
Botanic Garden (Belfast) 47
Boulée, Étienne-Louis
149, 162

Bourgeois, Sir Peter Francis
43
box construction 102
Bramante, Donato 228
Bremmer, H P 79
brick making 4, 16, 39, 40,
319, 359
brick-built architecture
326–9
bridge, Salginatobel 101–4
Brinkman, Johannes 347
British Museum (London)
23, 139, 140, 157
Bronze Age 296
Browning, Robert 110
Brunel, Isambard Kingdom
8, 45
Brunelleschi, Filippo 28,
218–20, 227–8, 360–1
Brutalism 9
Buckminster Fuller, Richard
347
Buddhism 7, 127–8, 137,
207–8, 216, 224, 305
Buffalo (New York State)
14, 61–5
Bukhara (Uzbekistan)
190–2, 256
Buontalenti, Bernardo 278
Burj Khalifa (Dubai) 292
Burlington, Lord 138
Burnham, Daniel 54, 61,
64–8, 70
Burton, Decimus 7, 185, 316
Byzantine architecture 14,
172, 289
Byzantine Empire 178

C
Cabala 125–6
Ca' d'Oro (Venice) 220, 310,
315–8
Ca' d'Oro Building
(Glasgow) 49
Cahill, John 287
Cairo 188, 293
Callimachus 19, 23
Cambodia 168, 206–9
Cambyses II, King of Persia
257
Campbell, Colen 133
Campidoglio (Rome) 133
Canada 300, 336–40

Canova, Antonio 141
Canterbury Cathedral 193
Capri (Italy) 154–7
Caracalla, Emperor 266–7
Carrère & Hastings 337
Carthage 264, 373
Casa Malaparte (Capri)
154–7
Case Study Houses 88
cast-iron-fronted
architecture 48–9
Castle Howard 134
Castles
Crac des Chevaliers
(Syria) 111, 116–20
Himeji Castle (Japan)
321–5
Cathars 247–8
cathedrals
Chartres 168, 200–3
Durham 6, 168, 193,
196–200, 256
Florence 219
Catherine Palace (St
Petersburg) 356, 363–6
Cavendish Square (London)
282
Cerceau II, Jacques
Androuet du 277, 279
Chapel of St John (Tower of
London) 193–6
Chareau, Pierre 91–3, 347
Charles I, King 280
Charles V, Holy Roman
Emperor 230
Chartres Cathedral 168,
200–3
Château de Vaux-le-Vicomte
(Maincy) 35–9, 133
Chatsworth House
(Derbyshire) 38, 46, 132,
134
Cheney, Mamah 151
Chernikov, Yakov 347
Chicago 8, 54, 65
Chicago School 55, 74
China
Forbidden City (Beijing)
304–7
Great Wall of China
298–301
Hanging Temple (Shanxi
Province) 223–6

Choragic Monument of
Lysicrates (Athens) 115
Christ Church Spitalfields
(London) 244
Chrysler, Walter P 97
Chrysler Building (New
York) 94–8, 108
churches
Basilica of San Lorenzo
(Florence) 218–20, 226–30
Chapel of St John (Tower
of London) 193–6
Church of the Sacred
Heart (Prague) 243–6
Church of the
Transfiguration (Khizi,
Russia) 325–6
Frauenkirche (Dresden)
360–3
Hagia Sophia (Istanbul)
180–4
Notre Dame du Haut
(Ronchamp) 246–50
Rock-cut churches
(Lalibela) 209–12
Sagrada Família
(Barcelona) 237–40, 169
St Catherine's Monastery
(Sinai) 177–80
Santa Maria dei Miracoli
(Venice) 221–3
see also cathedrals
Ciano, Galeazzo 155
The Circus (Bath) 280–6
cities 4, 251
Citrohan House (mass-
produced) 100
Clairvaux, Cistercian
monastery of 314
classicism 30, 33–4, 44, 58,
76, 147–9, 155, 222–3,
228, 234, 246, 334, 340,
367, 371
Clement VII, Pope 229–30
Coalbrookdale (Shropshire)
40
Colonnade Street (Palmyra)
267–71
Colosseum (Rome) 29–30,
50, 222, 285
Colvin, Howard 132
Comares Palace (Alhambra)
121

concrete
reinforced 8, 14, 84, 87,
101–4, 108, 153, 287, 310
Roman use of 25–7
concrete-frame construction
100
Confuscius 299
Constantine, Emperor 182
Constantinople see Istanbul
Constructivism 9–10, 104–6
Contarini, Marino 315–7
Cook, Captain James 144–5
Cook, Peter 347
Coombs, Robert 248–9
Copenhagen 141–3
Corinthian order 19–24,
269, 285
Court of the Lions
(Alhambra) 112, 122–3
Covent Garden Piazza
(London) 255, 256
Crac des Chevaliers (Syria)
111, 116–20
Cranston Tearoom
(Glasgow) 332
Cressing Temple (Essex)
311–5
Crusades 3, 116–9, 184
Crystal Palace (London)
46–7
Ctesiphon (Iraq) 301–4, 371
Cubism 98, 249
Cultural Revolution 224, 305
cultural terrorism 252, 375
Cuthbert, St 196, 198
Cyrus the Great, King of
Persia 175, 256–7
Czech Republic 243–6

D

Dalsace, Dr 92
Damascus (Syria) 261, 267
Dance, George 41–2
Dante Alighieri 28
Daoism 224
Darby III, Abraham 40
Darius I King of Persia
257–60
Darmstadt Artists' Colony
72
David, King 173–4
Davies, Colin 344
De Stijl 78–80, 105

Dee, John 124–6
Delhi order 112
della Porta, Giacomo 361
democratic architecture
158, 160
demolition 317, 369
Denmark 141–3
Desenfans, Noel 43
Dessau (Germany) 81–5
Dhaka (Bangladesh) 160–3
Dickens, Charles 368
Ditherington Flax Mill
(Shrewsbury) 14, 39–41
Djenné (Mali) 168, 240–3
Djoser, Pharaoh 17
Doesburg, Theo van 78
Doge's Palace (Venice) 316
Dome of the Rock
(Jerusalem) 168, 184–7,
197, 236
domes 7, 27, 44, 48, 183–4,
192, 234, 325–6, 360–2
Doric order 20–4, 30, 44,
112, 171, 222, 246, 285,
367, 369, 378
Dresden 74, 356, 360–3, 374
Dubai 292–3
Dudok, Willem Marinus 328
Dulwich College (London) 43
Dulwich Picture Gallery
(London) 41–5
Dunnett, James 166, 402
Durham Cathedral 6, 168,
193, 196–200, 256

E

E. V. Haughwout Building
(New York) 48–51
Easton Neston
(Northamptonshire) 38,
131–4
ecological design 352
Edward VII, King 146
Egypt 13, 15, 169, 291, 293
Great Pyramid at Giza 13,
15–6, 291, 293–8
Mortuary Temple of
Hatshepsut (Luxor) 20,
169–72
St Catherine's Monastery
(Sinai) 177–80
Step Pyramid of Djoser
(Saqqara) 15–8

Elagabalus, Emperor 266
Elizabeth, Empress of
    Russia 363–6
Elizabeth I, Queen 124
Elizabeth II, Queen 288
Elizabethan architecture 58,
    124, 134, 334
Ellis, Peter 51–4
Empire State Building (New
    York) 96
engineering 6–9
England see United
    Kingdom
Equitable Life Assurance
    Building (New York) 70
Esfahani, Ostad Ali Akbar
    231–2
Ethiopia168, 175, 209–12
Euclid 124
Euston Arch (London) 356,
    366–9
Euston Station (London)
    366–9
Exeter College Chapel
    (Oxford) 59–60
Exhibition of the Industry
    of All Nations (New
    York) 51
Exposition Internationale
    des Arts Décoratifs et
    Industriels Modernes
    (Paris) 105
Expressionism 238
exterior display of services
    345

F

Fallingwater (Edgar J
    Kaufmann House)
    (Pennsylvania) 150–4
Farmhouse (Shengana,
    Bhutan) 135–7
Farnsworth House (Plano,
    Illinois) 310, 340–3, 348
Fascism 154–7
Ficino, Marsilio 125
Finland 157–60
Firozkoh 205
Flagg, Ernest 70
Flatiron Building (New
    York) 64–8, 71
Florence
    Basilica of San Lorenzo

218–20
New Sacristy (San
    Lorenzo, Florence)
    226–30
Palazzo Rucellai
    (Florence) 27–32
urban architecture 29, 348
Fludd, Robert 125
flying buttresses 199–201
Fontana, Domenico 361
Forbidden City (Beijing)
    304–7
Foreign Office (London)
    8, 60
Foster, Norman 344
Fouquet, Nicolas 36, 38
Fox, Charles 367
France
    barns 314
    Chartres Cathedral 200–3
    Château de Vaux-le-
        Vicomte (Maincy)
        35–9, 133
    Notre Dame du Haut
        (Ronchamp) 169,
        246–50
    see also Paris
Frauenkirche (Dresden) 356,
    360–3
Freemasonry 284
Freud, Ernst 88
Freud, Lucien 88
Freud, Sigmund 88, 332
Froebel, Friedrich 153
Fuller Building (Chicago) 65
Fun Palace 347
functionalism 53, 75–6, 93,
    149, 289, 348

G

Gabo, Naum 105
Gallienus, Emperor 268
Gardner's Warehouse
    (Glasgow) 49, 51, 53
Garrard, Margaret 283
Gate of all Nations
    (Persepolis) 259
Gaudí, Antoni 169, 237–40
Gay, Robert 281, 283
Gaynor, John P 48, 50
Gehry, Frank 310, 336–40
Genghis Khan 205, 373
George V, King 146

George Fuller Construction
    Company 65
Georgian architecture 280–6
Germany
    AEG Turbine Hall (Berlin)
        72–4
    Altes Museum (Berlin)
        138–40, 148, 157
    The Bauhaus (Dessau)
        81–5
    Frauenkirche (Dresden)
        356, 360–3
Ghiyath al-Din 206
Ghorid Empire 204–6
Gilbert, Cass 69–72
Gilgamesh 5, 255–6
Ginzburg, Moisei 89
Giotto di Bondone 28
Gisborne (New Zealand)
    144–6
Glasgow 49–53, 332
glass cladding 109, 337
Glyptothek (Munich) 139
Go-Mizunoo, Ex-emperor
    130
Goldfinger, Ernö 10, 163–6
Gómez de Mora, Juan 279
Gothic architecture 49,
    58–9, 158, 193, 198, 223,
    314, 317
Gothic Revival architecture
    57–60, 147, 238, 318,
    333–4, 367
Graham, Ernest R 70
Granada (Spain) 120–3
Grand Canal (Venice) 220,
    223, 315–8
Grand Central Terminal
    (New York) 337
Grand Tour 42
Great Exhibition (1851) 46
Great Mosque (Djenné,
    Mali) 168, 240–3
Great Mosque of Samarra
    168, 187–90
Great Pyramid at Giza 13,
    15–6, 291, 293–8
Great Wall of China
    298–301
Greece, ancient 13–4
    Temple of Apollo
        Epicurius (Bassae) 13,
        19–24

Tower of the Winds (Athens) 113–6
see also classicism; Corinthian order; Doric order; Ionian order
Greek Revival architecture 57, 142, 356, 367
green architecture 11, 311, 350
Grimshaw, Nicholas 344
Gropius, Walter 72–3, 81–5
Grosvenor Square (London) 282, 284
Guaranty Building (Buffalo, New York) 14, 61–5
Guernica (Spain) 373
Guggenheim Museum (Bilbao) 337–8
Gundulf, Bishop of Rochester 193–5

H

Hachijo Toshihito, Prince 126–31
Hadrian, Emperor 25–6
Haeinsa Temple (South Korea) 215–8
Hagia Sophia (Istanbul) 168, 180–4, 236, 361
Hall, Peter 288
Hall, Todd & Littlemore 288
Hanging Gardens of Babylon 358
Hanging Temple (Shanxi Province) 223–6
Harappa (Pakistan) 254
Hardwick, Philip 366–9
Harriman House (New York State) 336
Hastings, Battle of 193, 195
Hathor 171
Hatra (Iraq) 2, 252, 261, 267, 302, 357, 370–5
Hatshepsut, Pharaoh 20, 169–72
Hattin, Battle of 119
Hawksmoor, Nicholas 131–4, 244
Heathcote (Ilkley, Yorkshire) 336
Henri IV, King of France 279–80
Henrietta Maria, Queen 280

Hermes 125, 371
Herod I, King 173, 175, 184
Herrara, Juan de 255
high-rise design 7–9, 14, 50, 61
see also skyscrapers
High Tech architecture 9, 92–3, 310, 343–9
Hilversum Town Hall (Netherlands) 328
Himeji Castle (Japan) 321–5
Hinduism 208, 213
Hitler, Adolf 374
Holkham Hall (Norfolk) 138
Holy Sepulchre (Jerusalem) 181, 185, 236
Honeyman, John 49
Hopkins, Michael 343–8
Hopkins, Patti 344
Hopkins House (Hampstead, London) 343–8
Hormizd IV, King 302
Hradcany Castle (Prague) 246
Hulagu Khan 187
Hussein, Saddam 358

I

iconoclasm 375
Imam Mosque (Isfahan) 231–4
Imhotep 17
Imperial Hotel (Tokyo) 152
Impressionism 72
Inca civilization 252, 274–7
India
Sri Ranganathaswamy Temple (Srirangam) 168, 212–5
Taj Mahal (Agra) 168, 234–7
Indus Valley civilization 254
Industrial Modernism 153
industrial production 88, 346
Industrial Revolution 8, 14, 40, 55, 108
International Style 77, 196, 335
Les Invalides (Paris) 362
Inwood, William and Henry 116

Ionic order 20–4, 30–1, 37, 113, 140, 222, 245–6, 285, 378
Iran
Imam Mosque (Isfahan) 231–4
Persepolis 256–60
Iraq
Hatra 2, 252, 261–2, 302, 357, 370–5
Ishtar Gate (Babylon) 357–60
Malwiya (Sumarra) 187–90
Taq-i Kisra (Ctesiphon) 301–4
Uruk 5, 16, 189, 192, 253–6
Iron Bridge (Coalbrookdale) 40
iron-framed buildings 14, 39, 48–51
Ise Grand Shrine (Japan) 356
Isfahan (Iran) 231–4
Ishtar Gate (Babylon) 357–60
Isidorus of Miletus 180–1
Islamic architecture 3, 184–7, 197, 203–6, 231–4, 240–3
Islamic State (IS) 252, 269, 372, 375
Isma'il ibn Ahmad, amir of Transoxiana 190
Istanbul 178, 180–6
Italian Futurism 105, 347
Italy
Ca' d'Oro (Venice) 220, 310, 315–8
Casa Malaparte (Capri) 154–7
Santa Maria dei Miracoli (Venice) 221–3
Villa Almerico Capra (La Rotonda) (Vicenza) 32–5
see also Florence; Roman architecture

J

Jacobean architecture 124, 334
James I, King 124
James, Henry 69

Janggyeong Panjeon (South Korea) 215–8
Japan
architecture 6, 93, 126–31, 152, 321–5
Himeji Castle 321–5
Old Shoin and the New Palace (Kyoto) 126–31
Jayavarman VII, King 207–9
Jekyll, Gertrude 334–6
Jericho 254
Jerusalem
Dome of the Rock 168, 184–7, 197, 236
Temple of Solomon 172–6, 181, 184, 186, 201, 212, 277
Jethro 177
John the Divine, St 176, 179
Jones, Inigo 281
Jordan, Petra 260–3
Justinian I, Emperor 178–82

K

Kahn, Louis 10, 160–3
Kaplicky, Jan 344
Katsura Imperial Palace (Kyoto) 126–131
Keene, Henry 116
Kent, William 138
Kew Gardens (London) 45–8
Khmer Empire 206
Khosrau I, King 302
Khosrau II, King 302
Khufu, Pharaoh 293–4, 298
Klenze, Leo von 139
Klimt, Gustav 299, 331–2
Knights of St John of Jerusalem (Hospitallers) 119
Knights Templar 119, 185, 202, 248, 311
Koenig, Pierre 88
Koy Konoboro, King 242
Krishna 214
Kuwabara Payne McKenna Blumberg Architects 337
Kyoto (Japan) 126–31

L

Labrouste, Henri 48
Lake Shore Apartments (Chicago) 108
Lalibela (Ethiopia) 209–12

Lalibela, King of Ethiopia 209–12
Lamb, William F 96
Lanyon, Charles 47
Lao-Tzu 224
Lasdun, Denys 10
Laugier, Marc-Antoine 19
Lawrence, T E 116
Le Brun, Charles 35–9
Le Corbusier 10, 72, 79, 89–94, 98–100, 164, 169, 245–50, 289, 335–6, 346–7
Le Notre, André 35–9
Le Vau, Louis 35–9
Ledoux, Claude Nicolas 112, 149, 162
Leningrad, Siege of 366
Leominster, Sir William Fermor, Lord 132–4
Leptis Magna (Libya) 264–7
Libera, Adalberto 154–7
Libya 264–7
Lily House (Chatsworth) 46
limestone 16, 18, 22, 67, 217, 281–2, 294–6, 337
Lincoln, Mrs 51
Lincoln's Inn Fields (London) 42
Linke Wienzeile apartment blocks (Vienna) 332
Lion Gate (Mycenae) 23
Lippi, Filippo 31
Liverpool 51–4
Lloyd's of London 9, 348–50
Lombardo, Pietro 221–3
London
Chapel of St John (Tower of London) 193–6
Dulwich Picture Gallery 41–5
Euston Arch 366–9
Hopkins House (Hampstead) 343–8
Lloyd's of London 348–50
The Palm House (Kew Gardens) 45–8
St Pancras Station 55–61
squares 153, 185–6, 195, 197, 199, 219, 284, 325
Trellick Tower 163–6
Longhena, Baldassare 361
Loos, Adolf 74–7, 79, 86, 330, 335, 369

Los Angeles 85–8
Louis XIV, King of France 36, 365
Lovell, Philip 85–8
Lovell Health House (Los Angeles) 85–8
Lübeck 374
Ludwig I, King of Bavaria 139
Luther, Martin 230
Lutyens, Sir Edwin 112, 147, 333–6
Luxor (Egypt) 20, 169–72
Lynn, Jack 89

M

machine-age architecture 348–50
Machu Picchu (Peru) 374–7
McKim, Mead and White 147–337
Mackintosh, Charles Rennie 332
Macrinus 266
magic 123–6
Maillart, Robert 101–4
Maincy (France) 35
Maison Carée (Nimes) 29
La Maison de Verre (Paris) 91–4
Majolikahaus (Vienna) 299, 330–3
Malaparte, Curzio 154–7
Mali 240–3
Malwiya (Sumarra, Iraq) 187–90
Manaus (Brazil) 286
Manchuria 373
mandalas 168, 212
Manetho 17
Mansart, Jules Hardouin 362
Maoris 144–6
Marathon, Battle of 258
marble-clad buildings 223, 317
Marshall, Benyon and Bage's Flax Mill (Ditherington, Shrewsbury) 39–41
Marshcourt (Stockbridge, Hampshire) 333–6
Martorell, Joan 238
Mary, Queen 146
mastabas 18
Matilda, Empress 311

Mausoleum of Isma'il
Samani (Bukhara) 190–3,
256
Maya civilization 252, 271–4
Mecca 188, 233, 242
Medici, Cosimo de' 31, 138
Medici, Giuliano de' 229
Medici, Lorenzo de' (the
Magnificent) 31, 229
Medici, Nannina de' 31
Medici, Piero di Cosimo
de' 31
Medici family 18, 227, 230
Medina 188, 232
Melnikov House (Moscow)
104–7
Melnikov, Konstantin 104–7
Mendelsohn, Erich 86
Menelik I, Emperor of
Ethiopia 175, 211
Merchants' Exchange
(Philadelphia) 116
Mesopotamia 5, 13, 15–6,
191–2, 200, 252–4, 261,
302, 359, 370
Metropolitan Life Insurance
Tower (New York) 70
Mexico 252, 271–4
Meyer, Hannes 84
Michelangelo Buonarroti
133, 226–30, 361
Midland Bank Headquarters
(London) 147
Midland Grand Hotel
(London) 55–60, 158
Mies van der Rohe, Ludwig
14, 72, 84, 107–10, 246,
340–3
Milinis, Ignaty 89
military architecture 116,
194, 321
minarets
Malwiya (Sumarra, Iraq)
187–90
Minaret of Jam
(Afghanistan) 168, 203–6
Ming dynasty 225, 298–301
minimalism 42, 54, 75, 181
Modernism 53, 61–3, 72–3,
75, 80, 84, 87, 90–2, 100,
105, 147–9, 152–3, 157–8,
245, 247, 289, 327–9,
335–6, 345

Mohenjo-daro (Pakistan) 254
Monadnock Building
(Chicago) 65
Mondrian, Piet 78–80
Mongols 187, 205, 216,
373, 375
Moors 120
Morris, William 334
Mortuary Temple of
Hatshepsut (Luxor) 20,
169–72
Moscow 89–90, 104–7,
164, 364
Moses 4, 125, 177–8
mosques
Great Mosque (Djenné,
Mali) 168, 240–3
Hagia Sophia (Istanbul)
168, 180–4, 236, 361
Imam Mosque (Isfahan)
321–4
Mosque of Ahmed ibn
Tulun (Cairo) 188
Mosque of Kairouan
(Tunisia) 188
Mosque of the Nine
Domes (Balkh) 192
Mount Gaya (South Korea)
215–8
Mount Hengshan (China)
223
Mount Stewart (County
Down) 116
mud-brick buildings 256
Mughal architecture 234–7
Muhammad V, Sultan of
Granada 122
Muhammad, the Prophet
187–8
Müller House (Prague) 77
Mumtaz Mahal 235–6
Municipal Museum (The
Hague) 298
Munstead Wood (Surrey)
329
murder holes 118, 324
museums and galleries
Altes Museum (Berlin)
138–40, 148, 157
Art Gallery of Ontario
(Toronto) 310, 336–40
Dulwich Picture Gallery
(London) 41–5

Thorvaldsen Museum
(Copenhagen) 141–3
Mussolini, Benito 154–5
al-Mu'tasim, Caliph 189
Mycenae 23
Myers, Barton 337

N
Nabateans 260–1
Napoleon LeBrun & Sons 70
Naram-Sin, King of Uruk 5
Narkomfin Apartment
Block (Moscow) 87–91,
164
Nashdom (Taplow,
Buckinghamshire) 336
National Gallery (London)
139
National Parliament
Building (Dhaka) 160–3
National Theatre (London)
10
Nazis 84–5, 106, 373
Nebuchadnezzar II, King of
Babylon 5, 255, 257–8, 260
Nemours, Giuliano de'
Medici, Duke of 229
neo-classicism 57, 91, 104–6,
115, 142, 331, 336
neo-Georgian 336
neo-Gothic 169, 238, 331,
344–5
neo-Plasticism 78–9
neo-Renaissance 336
neo-Romanesque 329, 331
The Netherlands
Beurs van Berlage
(Amsterdam) 326–9
Rietveld Schröder House
(Utrecht) 77–81
Neutra, Richard 85–8
New Delhi 112
New Forum (Leptis Magna)
264–7
New Jerusalem 176, 179,
180, 195–6, 198, 209–210,
212
New Sacristy (San Lorenzo,
Florence) 226–30
New York 8
Chrysler Building 94–8
E. V. Haughwout Building
48–51

Flatiron Building 64–8
Public Library 337
Seagram Building 107–10
Woolworth Building 69–72
New Zealand 144–6
Newgrange (County Meath) 189
Nile, River 57, 169, 171, 295
Nimrud 2, 374–5
Nineveh 2, 374–5
al-Nisaburi, Ali ibn Ibrahim 203–6
Noel, Miriam 152
nonce orders 112
Normans 194, 197
Notre Dame du Haut (Ronchamp) 169, 246–50
numbers, in architecture 35, 124, 186, 188, 233, 254, 256, 263

O

oak-framed construction 311–5
occult 124–6
Ogedei 205
Olbrich, Joseph Maria 75, 331
Old Shoin and the New Palace (Kyoto) 126–131
Olympieion (Athens) 22
Oriel Chambers (Liverpool) 51–4
ornament in architecture 19, 30, 48, 50, 55–6, 58, 60–2, 71, 73–6, 111, 135, 190, 265, 329–33
Osborne, Lady Sophia 132
Osiris 172
Ospedale degli Innocenti (Florence) 219
Otis elevators 51, 67
Ottoman Empire 315
Ozenfant, Amédée 98

P

Paddington Station (London) 8
Paestum 22
Paikea 146
Pakal, King of Palenque 271–4

Pakistan 162, 204, 254
palaces
Alhambra (Granada) 112, 120–3
Ca' d'Oro (Venice) 220, 310, 315–8
Catherine Palace (St Petersburg) 356, 363–6
Château de Vaux-le-Vicomte (Maincy) 35–9, 133
Forbidden City (Beijing) 304–7
Katsura Imperial Palace (Kyoto) 126–31
Palace of Westminster (London) 58
Palais du Louvre (Paris) 36
Palais de Versailles 36, 38, 365
Palazzo Barberini (Rome) 37
Palazzo Dario (Venice) 223
Palazzo Rucellai (Florence) 27–32
Palenque (Mexico) 271–4
Palladian architecture 37, 281–2
Palladio, Andrea 32–5, 37, 98
The Palm House (Kew Gardens) 45–8
Palmerston, Lord 60
Palmyra (Syria) 252, 267–71, 370–4
Pantheon (Rome) 14, 24–7, 37, 140, 161, 181, 220, 227, 361–2
Parçay-Meslay (Indre-et-Loire) 314
Paris
La Maison de Verre 91–4
Place des Vosges 252, 277–80
Villa Savoye 92, 98–100, 245, 247
Park Hill Flats (Sheffield) 89
Parthenon (Athens) 21, 22, 68
Parthian Empire 301–2, 370–1
Pausanius 21
Paxton, Joseph 45–7
Pearson and Darling 336

Pennsylvania Station (New York) 147
Pergamon Museum (Berlin) 356–7, 360
Perrault, Dominique 113
Perrières, Benedictine priory 314
Persepolis (Iran) 256–60
Persian architecture 7, 231, 256–60, 358
Peru 252, 274–7
Peter the Great, Tsar 363–4
Petra (Jordan) 252, 260–3, 267, 370
Petrie, William Flinders 296
Pevsner, Nikolaus 53
Phigalia 21
Philip II, King of Spain 280
Philip Lovell Beach House (Newport Beach, California) 87
Phoenicians 264
Piano, Renzo 9, 344, 349
Piazza Grande (Livorno) 278, 281
Pico della Mirandola, Giovanni 125
pilotis 90, 99–100
Place Dauphine (Paris) 279
Place des Vosges (Paris) 252, 277–80
Plano (Illinois) 310, 340–3
Plato 24
Plaza Mayor (Madrid) 279
Plecnik, Joze 243–6
Plug-In City 347
Plutarch, Francesco 28
Pompidou Centre (Paris) 9, 349
Post Office Savings Bank (Vienna) 245, 333
Post-Modernism 336
Prague 77, 243–6
Prairie School 151, 153
Price, Cedric 347
Pritchard, Thomas Farnolls 40
Processional Way (Babylon) 259, 357–9
Prudential Building (Buffalo) see Guaranty Building
Pu Yi, Emperor 305

public housing 89, 247
Pugin, A W N 10, 58–9, 62, 334–5
pyramids 12
  Great Pyramid at Giza 293–8
  Pyramid of Khafre 294
  Pyramid of Menkaure 294
  Step Pyramid of Djoser 15–18
  Temple of the Inscriptions (Palenque) 272

Q

Qin Shi Huang, Emperor 299
Queen Anne Revival 334
Queen Square (Bath) 282–6

R

Ra 295
Radcliffe Observatory (Oxford) 116
Rastrelli, Bartolomeo 363–6
Raymond of Tripoli, Count 119
Realistic Manifesto (Gabo) 105
reconstruction 33, 50, 117, 139, 176, 178, 184, 187, 197, 201, 270–1, 303–4, 358, 369
Reed & Stem 337
Reliance Building (Chicago) 54, 65
Reliance Controls factory (Swindon) 344
religious intolerance 252, 373
Rembrandt 43
Renaissance architecture 7, 27–32, 218–23, 226–30
renewable energy 310, 351
Revett, Nicholas 115
Richardson, Henry Hobson 329
Rietveld, Gerrit 77–81, 335
Rietveld Schröder House (Utrecht) 71–81
Robin Hood Gardens (Poplar, London) 89
Rock-cut churches (Lalibela) 209–12

Rococo architecture 57
Rogers, Richard 9, 344, 348–50
Roman architecture 4, 7
  influence of 27–32, 219–23
  Pantheon (Rome) 14, 24–7, 37, 161, 181, 220, 227, 361–2
  see also classicism
Romanesque architecture 193, 196, 198–9, 256, 329
Ronchamp (France) 246–50, 289
Rookery (Chicago) 54
Root, John Wellborn 54, 61, 65
Rosellino, Bernardo 28
La Rotonda *see* Villa Almerico Capra
Rotterdam 347, 373
Royal Albert Bridge (Saltash) 45
Royal Crescent (Bath) 384–5
Royal Saltworks (Arc-et-Senans) 112
Rozak House (Northern Territory, Australia) 350–4
Rubens, Peter Paul 43
Rucellai, Bernardo 31
Rucellai, Giovanni 31
Rushton Triangular Lodge (Northamptonshire) 112, 123–6
Ruskin, John 55–6, 223, 317–8, 334
Russia
  Catherine Palace (St Petersburg) 356, 363–6
  Church of the Transfiguration (Khizi) 325–6
  Melnikov House (Moscow) 104–7
  Narkomfin Apartment Block (Moscow) 87–91, 164

S

sacred architecture 167
Sagrada Família, Temple Expiatori de la (Barcelona) 169, 237–40
St Albans, Henry Jermyn, Earl of 280

St Catherine's Monastery (Sinai) 177–80
St Denis (Paris) 199
St George's Church (Lalibela) 211
St George's Hall (Liverpool) 158
St James Square (London) 280
St Pancras New Church (London) 116
St Pancras Station (London) 14, 55–61, 347
St Paul's Cathedral (London) 234, 362
St Peter's Basilica (Rome) 181
St Petersburg (Russia) 105, 356, 363–6
Sainte-Chapelle (Paris) 60
Saladin 119
Salginatobel Bridge (Schiers, Switzerland) 101–4
Samurai 128
Sanctuary of Bel (Palmyra) 269–70
Santa Maria dei Miracoli (Venice) 221–3
Santa Maria della Salute (Venice) 361
Santa Maria Novella (Florence) 31
Sant'Elia, Antonio 347
Saqqara 13, 15–8, 293
Sassanian Empire 301–3
Säynätsalo Town Hall (Finland) 157–8
Schiers (Switzerland) 101–4
Schindler, Rudolph 85–8
Schindler House (Los Angeles) 86–7
Schinkel, Karl Friedrich 138–40
Schröder-Schräder, Truus 78–9
Scott, George Gilbert 8, 55–61, 158
Seagram Building (New York) 107–110
Secession Building (Vienna) 331–2
Second World War 108, 147, 264, 356, 373–4
Seleucid Empire 370

Semper, Gottfried 327, 330
Senenmut 169–72
Sennacherib, King of
    Assyria 373
Septimius Odainat 268
Septimius Severus, Emperor
    264–6, 302, 370
'Seven Sisters' (Moscow) 91
Seven Wonders of the
    World 293
Severan dynasty 266–7
Severance, H Craig 95
Severus Alexander, Emperor
    266–7
Shah Jahan 235, 237
Shakespeare, William 124
Shapur I, King 302, 372
Shaykh-i Bahai 231–2
Sheba, Queen of 175
Sheshonq I, King of Egypt
    174
Shibam (Yemen) 310, 318–21
Shikibu, Murasaki 127
Shrieve & Lamb 95
shrines
    Dome of the Rock
        (Jerusalem) 184–7
    see also tombs
Shugborough Park
    (Staffordshire) 116
Siem Reap (Cambodia)
    206–9
Silk Road 193, 203
Sinai 177–80
Singer Building (New
    York) 70
Sistine Chapel (Rome) 230
skyscrapers 14, 39
    Chrysler Building (New
        York) 94–8
    Flatiron Building (New
        York) 64–8
    Guaranty Building
        (Buffalo) 61–4
    Seagram Building (New
        York) 107–10
    Woolworth Building
        (New York) 69–72
Smirke, Robert 139
Smith, Adrian 293
Smith, Ivor 89
Smithson, Peter and Alison

Smollett, Tobias 285
Soane, Sir John 41–5, 57,
    138, 141
Socialist Realism 106
Solomon, King 23, 172–6,
    181, 201, 211
SOM 293
South Korea 215–8
Soviet architecture 9, 91,
    104–7
Spain
    Alhambra (Granada) 112,
        120–3
    Sagrada Família
        (Barcelona) 169, 237–40
Spanish Civil War 239, 373
Spenser, Edmund 126
Sphinx 208, 294
Sri Ranganathaswamy
    Temple (Srirangam, India)
    168, 212–5
Stahl House (Los Angeles)
    88
stained glass 3, 201–2, 249
Stalin, Josef 91, 106, 166
Stam, Mart 347
Stanton Drew stone circle
    282–4
steel-framed buildings 71, 88
steelwork, structural 8,
    348–50
Steiner, Lilly 74
Steiner House (Vienna) 74–7
Step Pyramid (Saqqara) 13,
    15–8, 295
Stephanos of Aila 178–9
Stephenson, George and
    Robert 367
Stieglitz, Alfred 68
Stockbridge (Hampshire)
    333–6
Stockholm Public Library
    147–50
Stonehenge 15–6, 59, 282–4
Strickland, William 116
Strutt, William 41
Stuart, James 115–6
Subirachs, Josep Maria 240
Suckert, Kurt Erich see
    Malaparte, Curzio
Sullivan, Louis 61–4, 74–5,
    97, 329–30, 332
Sumarra (Iraq) 168, 187–90

Sumeria 16, 189, 254
sustainable architecture 11,
    243, 310, 351–4, 319
Sweden 147–50
Switzerland 101–4
Sydney Opera House 286–9
Syria
    Crac des Chevaliers 111,
        116–20
    Palmyra 252, 267–71, 370–4

T
Taglioni, Maria 317
Taj Mahal (Agra) 234–7
Talman, William 132, 134
Taoism 224–6, 305–6
Taq-i Kisra (Ctesiphon)
    301–4
Tatlin, Vladimir 105
Team 4 344
Teitelbaum, Matthew 338
temples
    Bayon (Angkor Thom)
        168, 208–9
    Haeinsa Temple (South
        Korea) 215–8
    Hanging Temple (Shanxi
        Province) 223–6
    Machu Picchu 274–7
    Mortuary Temple of
        Hatshepsut (Luxor) 20,
        169–72
    Sri Ranganathaswamy
        Temple (Srirangam,
        India) 168, 212–5
    Temple of Aphaia
        (Aegina) 22
    Temple of Apollo
        Epicurus (Bassae) 19–24
    Temple of the Inscriptions
        (Palenque) 272–4
    Temple of Neptune
        (Paestum) 22
    Temple of Portunus
        (Rome) 29
    Temple of Solomon
        (Jerusalem) 23, 123,
        172–6, 181, 184, 186,
        188, 199, 201–2, 211–2,
        231–2, 277–9, 282,
        298, 357
    The Treasury (Khazneh)
        (Petra) 260–3